# SpringerBriefs in Mathematics

**SpringerBriefs in Mathematics** showcases expositions in all areas of mathematics and applied mathematics. Manuscripts presenting new results or a single new result in a classical field, new field, or an emerging topic, applications, or bridges between new results and already published works, are encouraged. The series is intended for mathematicians and applied mathematicians.

More information about this series at http://www.springer.com/series/10030

Eli Levin • Doron S. Lubinsky*

# Bounds and Asymptotics for Orthogonal Polynomials for Varying Weights

Springer

*Research supported by NSF grant DMS136208

Eli Levin
Department of Mathematics
Open University of Israel
Tel-Aviv, Israel

Doron S. Lubinsky
Mathematics
Georgia Institution of Technology
Atlanta, Georgia, USA

ISSN 2191-8198                    ISSN 2191-8201    (electronic)
SpringerBriefs in Mathematics
ISBN 978-3-319-72946-6          ISBN 978-3-319-72947-3    (eBook)
https://doi.org/10.1007/978-3-319-72947-3

Library of Congress Control Number: 2017963056

Mathematics Subject Classification: 42C05, 41A17, 30C15, 30E15, 31A15

Printed on acid-free paper

This Springer imprint is published by Springer Nature
The registered company is Springer International Publishing AG
The registered company address is: Gewerbestrasse 11, 6330 Cham, Switzerland

# Acknowledgement

The authors thank Annette Rohrs for her meticulous care in preparing the manuscript, and Danielle Walker and Donna Chernyk for their prompt and efficient handling of the process.

# Contents

# Contents

# Chapter 1
# Introduction

Let $\mu$ be a finite positive Borel measure with support on the real line containing infinitely many points, and all finite power moments $\int x^j d\mu(x)$, $j = 0, 1, 2, \ldots$. We may then define orthonormal polynomials $p_n(x)$ of degree $n$, $n = 0, 1, 2, \ldots$, satisfying

$$\int p_m(x) p_n(x) d\mu(x) = \delta_{mn}.$$

The asymptotic behavior of $p_n(x)$ as $n \to \infty$ has been studied for over a century. Beginning around 1918 [47, 48] Szegő analyzed orthonormal polynomials for absolutely continuous measures supported on $[-1, 1]$, or the unit circle, motivated by connections to Hankel and Toeplitz matrices. Plancherel and Rotach in the late 1920s [41] considered the Hermite weight $\mu'(x) = e^{-x^2}$ on $(-\infty, \infty)$, in order to investigate convergence of orthonormal expansions in Hermite polynomials. Plancherel and Rotach applied the method of steepest descent to a contour integral representation of Hermite polynomials. The very precise asymptotics they established are now called Plancherel-Rotach type asymptotics, and continue to be studied for more general measures to this day.

Until the last three decades of the 20th century, there were very few techniques for investigating orthogonal polynomials for non-compactly supported measures. If the orthogonal polynomials admit a contour integral representation, or a simple second order differential equation, or have a generating function as in the case of Pollaczek polynomials, classical asymptotic methods are applicable. However even the rudiments of a general theory were lacking. It was Geza Freud and later Paul Nevai who in the 1970s began to consider general weights $e^{-2Q(x)}$ on $(-\infty, \infty)$, using extremal properties and approximation to develop weaker forms of asymptotics. Nevai and his students, William Bauldry, Stan Bonan, Rong Sheen, and Shing Whu-Jha, obtained precise asymptotics for weights like $\exp\left(-x^{2m}\right)$, where

© The Author(s) 2018
E. Levin, D.S. Lubinsky, *Bounds and Asymptotics for Orthogonal Polynomials for Varying Weights*, SpringerBriefs in Mathematics,
https://doi.org/10.1007/978-3-319-72947-3_1

$m$ is a positive integer, using a mixture of analyzing differential equations and recurrence relations. Paul Erdős provided valuable insights for the case where $\mu'(x)$ decreases faster than $e^{-|x|^{\alpha}}$ for all $\alpha > 0$. See the still very relevant 1986 survey paper of Nevai [40].

Potential theory with external fields provided a dramatic breakthrough in the 1980s. In landmark papers, E. A. Rakhmanov [42] and Mhaskar and Saff [36–38] showed how to analyze orthogonal and extremal polynomials for quite general weights of the form $e^{-2Q(x)}$ on the real line. A comprehensive and polished development of that theory appears in the celebrated monograph of Saff and Totik [44]. By combining that potential theory with older methods of orthogonal polynomials, such as explicit formulae for Bernstein-Szegő weights, many researchers in orthogonal polynomials were able to analyze asymptotics – including the present authors [25].

An alternative approach to asymptotics for orthogonal polynomials is to place hypotheses on the coefficients in their three term recurrence relation, rather than on the underlying measure or weight. Some model examples of this approach for non-compactly supported measures appear in [14, 15, 55, 57]. Yet another relevant link is to discrete measures associated with indeterminate moment problems, see for example [7].

A second revolution for the case of absolutely continuous weights, came with the Deift-Zhou method in [9, 11, 12]. They developed a steepest descent method for a matrix Riemann-Hilbert problem whose solution includes orthonormal polynomials, and which was first observed by Fokas, Its, and Kitaev. The dramatic ramifications of that method continue to be observed to this day. While it initially dealt primarily with analytic or piecewise analytic weights, it has been extended by McLaughlin and Miller using a $\bar{\partial}$ approximation [34, 35]. A distinguishing feature of results obtained via Riemann-Hilbert methods is that they hold globally, and are far more precise than any general results obtained using any other method. Because they were motivated by problems arising in random matrices, Riemann-Hilbert researchers usually considered varying rather than fixed measures. That brings us to the setting of this monograph, which is the varying weights case.

For $n \geq 1$, let $\mu_n$ be a finite positive Borel measure with support $\mathrm{supp}[\mu_n] \subset \mathbb{R}$, containing infinitely many points. Assume also that all power moments $\int x^j d\mu_n(x)$, $j = 0, 1, 2, \ldots$, are finite. We may define orthonormal polynomials

$$p_{n,m}(\mu_n, x) = \gamma_{n,m}(\mu_n) x^m + \cdots, \quad \gamma_{n,m}(\mu_n) > 0,$$

$m = 0, 1, 2, \ldots$, satisfying the orthonormality conditions

$$\int p_{n,k}(\mu_n, x) p_{n,\ell}(\mu_n, x) d\mu_n(x) = \delta_{k\ell}.$$

The $n$th reproducing kernel for $\mu_n$ is

$$K_n(\mu_n, x, y) = \sum_{k=0}^{n-1} p_{n,k}(\mu_n, x) p_{n,k}(\mu_n, y). \qquad (1.1)$$

We often abbreviate this as $K_n(x, y)$. The $n$th Christoffel function is

$$\lambda_n(\mu_n, x) = K_n(\mu_n, x, x)^{-1}.$$

The absolutely continuous case, where

$$\mu_n'(x) = e^{-2nQ_n(x)} \qquad (1.2)$$

and $\{Q_n\}$ are given functions, plays an important role in random matrices. In this case, we often use the notation $p_{n,k}(e^{-2nQ_n}, x)$, $\lambda_n(e^{-2nQ_n}, x)$, and so on. The canonical example is $Q_n(x) = x^2$ for $n \geq 1$. Szegő style asymptotics of the associated orthonormal polynomials have been investigated by many authors, with many of the most spectacular results obtained using the Deift-Zhou steepest descent method. In particular, in celebrated papers [11, 12], Deift, Kriecherbauer, McLaughlin, Venakides, and Zhou considered the case where all $Q_n = Q$, and $Q(x)$ is real analytic on the real line, and grows faster than $\log |x|$ as $|x| \to \infty$. They established uniform asymptotics for the associated orthonormal polynomials in all regions of the complex plane, as well as detailed asymptotics for associated quantities, with applications to universality limits for random matrices. This set the stage for treating a large array of varying weights, such as varying (and sometimes fixed) Jacobi or Laguerre weights - some of the references are [3, 18–23, 56].

In all the earlier Riemann-Hilbert papers, $Q$ was required to be analytic in a neighborhood of the real line, or piecewise analytic. As noted above, using the $\bar{\partial}$-method, McLaughlin and Miller [34, 35] relaxed the requirement of analyticity, and considered the case where $Q''$ satisfies a Lipschitz condition of order 1, together with some other conditions. In particular, the latter conditions are satisfied when $Q$ is strictly convex in the real line. They established asymptotics for $p_{n,n}$ and $p_{n,n-1}$ in all regions of the complex plane – including asymptotics inside and at the edge of the Mhaskar-Rakhmanov-Saff interval (or equivalently, the support of the equilibrium measure). One of our foci is to further relax their smoothness requirements on $Q$.

We shall need some concepts from the potential theory for external fields [44], to which we alluded above. Let $\Sigma$ be a closed set on the real line, and $e^{-Q}$ be an upper semi-continuous function on $\Sigma$ that is positive on a set of positive linear Lebesgue measure. If $\Sigma$ is unbounded, we assume that

$$\lim_{|x| \to \infty, x \in \Sigma} (Q(x) - \log |x|) = \infty.$$

Associated with $\Sigma$ and $Q$, we may consider the extremal problem

$$\inf_{\nu} \left( \int \int \log \frac{1}{|x-t|} d\nu(x) \, d\nu(t) + 2 \int Q \, d\nu \right),$$

where the inf is taken over all positive Borel measures $\nu$ with support in $\Sigma$ and $\nu(\Sigma) = 1$. The inf is attained by a unique equilibrium measure $\omega_Q$, with support $\text{supp}[\omega_Q]$, characterized by the following conditions: let

$$V^{\omega_Q}(z) = \int \log \frac{1}{|z-t|} d\omega_Q(t) \tag{1.3}$$

denote the logarithmic potential for $\omega_Q$. Then [44, Thm. I.1.3, p. 27]

$$V^{\omega_Q} + Q \geq F_Q \text{ q.e. on } \Sigma; \tag{1.4}$$

$$V^{\omega_Q} + Q = F_Q \text{ q.e. in supp}[\omega_Q]. \tag{1.5}$$

Here the number $F_Q$ is a constant, and q.e. stands for quasi everywhere, that is, except on a set of capacity 0. Notice that we are using $\omega_Q$ for the equilibrium measure, rather than the more standard $\mu_W$ or $\nu_W$, to avoid confusion with $\mu_n$ or $\nu_n$. We use

$$\sigma_Q(x) = \omega'_Q(x) \tag{1.6}$$

for the Radon-Nikodym derivative of $\omega_Q$. Sometimes we denote $V^{\omega_Q}$ by $V^{\sigma_Q}$.

While the Riemann-Hilbert methods yield the strongest results for smooth weights, techniques based on potential theory and Bernstein-Szegő weights allow one to treat more general weights. Indeed, this was the traditional approach for fixed exponential weights adopted in [25, 32, 42, 43, 51, 52]. These methods enabled one to establish asymptotics of the orthonormal polynomials in the complex plane away from the interval of orthogonality, but not usually pointwise asymptotics on the interval. The most general results for varying weights, using these types of tools, were obtained by V. Totik in his 1994 lecture notes [51, Thm. 14.2, p. 99; Thm. 14.4, p. 101]:

**Theorem A.** *For $n \geq 1$, let $e^{-2nQ_n}$ be a weight function on $[-1, 1]$, whose equilibrium measure $\omega_{Q_n}$ has support $[-1, 1]$. Assume that $\omega_{Q_n}$ is absolutely continuous, and its density $\sigma_{Q_n}$ satisfies*

$$\frac{1}{A}\left(1 - t^2\right)^{\beta_0} \leq \sigma_{Q_n}(t) \leq A\left(1 - t^2\right)^{\beta_1}, \quad t \in (-1, 1),$$

*where $\beta_1 > -1$, and $A, \beta_0, \beta_1$ are independent of n. Assume also that $\{\sigma_{Q_n}\}$ are uniformly equicontinuous in every compact subset of $(-1, 1)$.*

*(I) Then for any fixed integer k,*

$$p_{n,n+k}\left(e^{-2nQ_n}, x\right) e^{-nQ_n(x)}$$

$$-\sqrt{\frac{2}{\pi}} \frac{1}{\sqrt[4]{1-x^2}} \cos\left[\left(k+\frac{1}{2}\right) \arccos x + n\pi \int_x^1 \sigma_{Q_n} - \frac{\pi}{4}\right] \qquad (1.7)$$

*tends to 0 in $L_2[-1, 1]$ as $n \to \infty$.*

*(II) Uniformly for z in compact subsets of $\bar{\mathbb{C}} \setminus [-1, 1]$,*

$$p_{n,n+k}(z) = \frac{1+o(1)}{\sqrt{2\pi}} \left(z + \sqrt{z^2-1}\right)^{k+\frac{1}{2}} \left(z^2-1\right)^{-1/4}$$

$$\times \exp\left(nF_{Q_n} - n \int \log \frac{1}{z-t} \sigma_{Q_n}(t)\, dt\right). \qquad (1.8)$$

*Here $F_{Q_n}$ is the constant in (1.4) for $Q = Q_n$.*

We note that this is not the most general form of Totik's result, and both asymptotics above can be formulated in terms of Szegő functions and their arguments. Indeed, (1.7) is formulated in a different way in [51]. Moreover, for $Q_n(x) = |x|^\alpha$, $\alpha > 1$, all the constants and densities can be given explicit forms. It is also significant that the weights $\{e^{-nQ_n}\}$ are assumed to be supported on $[-1, 1]$. There are extra difficulties in establishing asymptotics when, for example, the interval of orthogonality is unbounded. Then one has to use restricted range inequalities, and often this requires extra hypotheses.

Another important asymptotic is that for Christoffel functions. One of Totik's celebrated results for asymptotics of Christoffel functions for varying weights is [52]:

**Theorem B.** *Let $e^{-Q}$ be a continuous nonnegative function on the set $\Sigma$, which is assumed to consist of finitely many intervals. If $\Sigma$ is unbounded, we assume also*

$$\lim_{|x| \to \infty, x \in \Sigma} Q(x) / \log |x| = \infty.$$

*Let J be a closed interval lying in the interior of $\text{supp}[\omega_Q]$, where $\omega_Q$ denotes the equilibrium measure for Q. Assume that $\omega_Q$ is absolutely continuous in a neighborhood of J, and that $\sigma_Q$ is continuous in that neighborhood. Then uniformly for $x \in J$,*

$$\lim_{n \to \infty} \frac{1}{n} \lambda_n \left(e^{-2nQ}, x\right) e^{2nQ(x)} = \sigma_Q(x). \qquad (1.9)$$

In particular, when $Q'$ satisfies a Lipschitz condition of some positive order in a neighborhood of $J$, then [44, p. 209] $\sigma_Q$ is continuous there, and hence we obtain asymptotics of Christoffel functions there. Note too that when $Q$ is convex in $\Sigma$, or $xQ'(x)$ is increasing there, then the support of $\omega_Q$ consists of at most finitely many intervals, with at most one interval per component of $\Sigma$ [44, p. 199, Thm. 1.10(c)]. We used Totik's result to establish universality results for varying weights in [27, p. 747, Thm. 1.1].

Our aim in this paper is especially to establish locally uniform versions of (1.7) in compact subsets of the Mhaskar-Rakhmanov-Saff interval, as well as global bounds on the orthonormal polynomials. We now define the class of weights that we shall use throughout this book:

**Definition 1.1.** For $n \geq 1$, let $I_n = (c_n, d_n)$, where $-\infty \leq c_n < d_n \leq \infty$. Assume that for some $r^* > 1$, $[-r^*, r^*] \subset I_n$, for all $n \geq 1$. Assume that

$$\mu_n'(x) = e^{-2nQ_n(x)}, \quad x \in I_n, \tag{1.10}$$

where (i) $Q_n(x) / \log(2 + |x|)$ has limit $\infty$ at $c_n+$ and $d_n-$.
(ii) $Q_n'$ is strictly increasing and continuous in $I_n$.
(iii) There exists $\alpha \in (0, 1)$, $C > 0$ such that for $n \geq 1$ and $x, y \in [-r^*, r^*]$,

$$\left| Q_n'(x) - Q_n'(y) \right| \leq C |x - y|^\alpha. \tag{1.11}$$

(iv) There exists $\alpha_1 \in \left( \frac{1}{2}, 1 \right)$, $C_1 > 0$, and an open neighborhood $I_0$ of 1 and $-1$, such that for $n \geq 1$ and $x, y \in I_n \cap I_0$,

$$\left| Q_n'(x) - Q_n'(y) \right| \leq C_1 |x - y|^{\alpha_1}. \tag{1.12}$$

(v) $[-1, 1]$ is the support of the equilibrium distribution $\omega_{Q_n}$ for $Q_n$.
Then we write $\{Q_n\} \in \mathcal{Q}$.

**Remarks.** (a) The convexity and smoothness assumptions can be replaced by implicit assumptions involving bounds and smoothness of the equilibrium distributions such as bounds and smoothness.
(b) The support condition (v) is equivalent to the Mhaskar-Rakhmanov-Saff equations

$$\frac{1}{\pi} \int_{-1}^{1} \frac{xQ_n'(x)}{\sqrt{1 - x^2}} \, dx = 1; \tag{1.13}$$

$$\frac{1}{\pi} \int_{-1}^{1} \frac{Q_n'(x)}{\sqrt{1 - x^2}} \, dx = 0. \tag{1.14}$$

(c) It may seem strange that we impose a stronger smoothness condition near $\pm 1$ than elsewhere. This is needed to bound the equilibrium density near the endpoints

of the Mhaskar-Rakhmanov-Saff interval, as we shall see in Chapter 3. We also show
there that something like this is needed to ensure uniform convergence of integrals
that arise there, such as

$$\int_0^1 \frac{\left| Q_n'(t) - Q_n'(1) \right|}{(1-t)^{3/2}} \, dt.$$

(d) The Lipschitz condition of order $\alpha_1 > \frac{1}{2}$ in (iv) can be weakened to

$$\left| Q_n'(x) - Q_n'(y) \right| \le C_1 \Omega \left( |x - y| \right), \quad x, y \in I_n \cap I_0,$$

where

$$\int_0^1 \frac{\Omega(t)}{t^{3/2}} \, dt < \infty.$$

Under this weaker condition, we can still prove all the results of the next chapter,
but with weaker error terms, no longer with $O(n^{-\tau})$, some $\tau > 0$.
(e) For notational convenience, we shall often assume that

$$\alpha \le \alpha_1 - \frac{1}{2}. \tag{1.15}$$

(f) The hypotheses force for some $C > 0$ independent of $n$,

$$Q_n'(-1) \le -C \text{ and } Q_n'(1) \ge C,$$

see Lemma 3.2 below. Then for some $t_n \in (-1, 1)$,

$$Q_n'(t_n) = 0 \tag{1.16}$$

and then the uniform Lipschitz condition gives

$$\sup_{n \ge 1} \sup_{t \in [-r^*, r^*]} \left| Q_n'(t) \right| < \infty. \tag{1.17}$$

We can then divide each $\mu_n' = e^{-2nQ_n}$ by a normalizing constant, and assume that
also

$$Q_n(t_n) = 0, \tag{1.18}$$

and hence

$$\sup_{n \ge 1} \sup_{t \in [-r^*, r^*]} \left| Q_n(t) \right| < \infty. \tag{1.19}$$

(g) The hypotheses of Definition 1.1 are satisfied if, for example, for $n \geq 1$, $Q_n(x) = c_n |x|^{\beta_n}$ with all $\{\beta_n\}$ lying in a fixed compact subset of $(1, \infty)$, and $\{c_n\}$ are chosen so that the equilibrium measures have support $[-1, 1]$.

Throughout $C, C_1, C_2, \ldots$ denote constants independent of $n, x, t$ and perhaps other specified parameters. The same symbol does not necessarily indicate the same constant in different occurrences. For sequences $\{x_n\}$ and $\{y_n\}$ of nonzero real numbers, we write $x_n \sim y_n$ if there exists $C > 1$ such that for $n \geq 1$,

$$C^{-1} \leq x_n/y_n \leq C.$$

We shall state some of our main results in the next chapter. The proofs of these will be distributed over Chapters 3 to 15. We shall discuss the organization in more detail in the next chapter.

# Chapter 2
# Statement of Main Results

We first state our uniform bounds on the orthonormal polynomials and related quantities:

**Theorem 2.1.** *Assume that* $\{Q_n\} \in \mathcal{Q}$ *and that for* $n, m \geq 1$, $p_{n,m}$ *is the orthonormal polynomial of degree* $m$ *for the weight* $e^{-2nQ_n}$ *on* $I_n$.
*(a) Let* $A > 0$. *For* $n \geq 1$, *and*

$$|n - m| \leq An^{1/3}, \tag{2.1}$$

*we have*

$$\sup_{x \in I_n} |p_{n,m}|(x) e^{-nQ_n(x)} \left[|1 - |x|| + n^{-2/3}\right]^{1/4} \sim 1. \tag{2.2}$$

*Moreover, uniformly in such* $m, n$,

$$\sup_{x \in I_n} |p_{n,m}|(x) e^{-nQ_n(x)} \sim n^{1/6}. \tag{2.3}$$

*(b) Let* $A > 0$. *Uniformly for* $n \geq 1$, $m$ *satisfying* (2.1), *and* $x \in I_n$ *satisfying* $|x| \leq 1 + An^{-2/3}$, *we have*

$$\lambda_m\left(e^{-2nQ_n}, x\right) e^{2nQ_n(x)} \sim \frac{1}{n} \max\{|1 - |x||, n^{-2/3}\}^{-1/2}. \tag{2.4}$$

*Moreover, uniformly for* $n \geq 1$, $m$ *satisfying* (2.1), *and* $x \in I_n$,

$$\lambda_m\left(e^{-2nQ_n}, x\right) e^{2nQ_n(x)} \geq C \frac{1}{n} \max\{|1 - |x||, n^{-2/3}\}^{-1/2}, \tag{2.5}$$

© The Author(s) 2018
E. Levin, D.S. Lubinsky, *Bounds and Asymptotics for Orthogonal Polynomials for Varying Weights*, SpringerBriefs in Mathematics,
https://doi.org/10.1007/978-3-319-72947-3_2

*(c) Let $\{x_{jn}\}$ denote the zeros of $p_{n,n}$, ordered as*

$$x_{nn} < x_{n-1,n} < \cdots < x_{1n}.$$

*Uniformly for $n \geq 1$ and $1 \leq j \leq n-1$,*

$$x_{jn} - x_{j+1,n} \sim \frac{1}{n} \max\{|1 - |x_{jn}||, n^{-2/3}\}^{-1/2}. \tag{2.6}$$

*Moreover,*

$$1 - \frac{C_1}{n^{2/3}} \leq x_{1n} \leq 1 + \frac{C_2}{n}, \tag{2.7}$$

*with a similar inequality for $x_{nn}$.*

*Proof.* (a) See Theorems 7.1 and 14.2(d).
(b) See Theorem 5.1.
(c) See Theorems 6.1 and 14.2(c).　　　　　　　　　　　　　　　　　　　□

**Remarks.** We believe that for the uniform bound (2.2) to hold, one really does need $Q'_n$ to satisfy a Lipschitz condition of order at least $\frac{1}{2}$ near $\pm 1$.

　　Next, we turn to asymptotics on the interval of orthogonality.

**Theorem 2.2.** *Assume that $\{Q_n\} \in \mathcal{Q}$. Let $\varepsilon \in (0, \frac{1}{3})$. For $n \geq 1$, let $\sigma_{Q_n}$ denote the density of the equilibrium measure for $Q_n$ on $[-1, 1]$. There exists $\tau > 0$ such that uniformly for $n \geq 1$ and $|x| \leq 1 - n^{-\tau}$, $\theta = \arccos x$, and for*

$$|m - n| \leq n^{1/3-\varepsilon}, \tag{2.8}$$

*we have (a)*

$$\sqrt{\frac{\pi}{2}} p_{n,m}(x) e^{-nQ_n(x)} \left(1 - x^2\right)^{1/4}$$

$$= \cos\left((m-n)\theta + n\pi \int_x^1 \sigma_{Q_n}(t)\, dt + \frac{\theta}{2} - \frac{\pi}{4}\right) + O(n^{-\tau}). \tag{2.9}$$

*(b)*

$$\frac{1}{n}\sqrt{\frac{\pi}{2}} p'_{n,m}(x) e^{-nQ_n(x)} \left(1 - x^2\right)^{1/4}$$

$$= \pi \sigma_{Q_n}(x) \sin\left((m-n)\theta + n\pi \int_x^1 \sigma_{Q_n}(t)\, dt + \frac{\theta}{2} - \frac{\pi}{4}\right)$$

$$+ Q'_n(x) \cos\left((m-n)\theta + n\pi \int_x^1 \sigma_{Q_n}(t)\, dt + \frac{\theta}{2} - \frac{\pi}{4}\right) + O(n^{-\tau}). \tag{2.10}$$

*(c)*

$$\frac{1}{n}\lambda_n^{-1}\left(e^{-2nQ_n}, x\right)e^{-2nQ_n(x)} = \sigma_{Q_n}(x) + O\left(n^{-\tau}\right). \tag{2.11}$$

*(d) Uniformly for j with* $|x_{jn}| \le 1 - n^{-\tau}$,

$$n\sigma_{Q_n}(x_{jn})(x_{jn} - x_{j+1,n}) = 1 + O\left(n^{-\tau}\right). \tag{2.12}$$

*Proof.* (a), (b) See Theorem 13.2(a), (b).
(c) See Theorem 13.3.
(d) See Theorem 13.5(b).                                                                                     □

**Remarks.** (a) We expect that one can prove the asymptotic (2.11) for the Christoffel function without assuming the extra Lipschitz condition (1.12) near ±1.

(b) We also expect that one can prove the asymptotic (2.9) assuming less smoothness on $\{Q_n\}$: instead of (1.11), assume equicontinuity of $\{\sigma_{Q_n}\}$ in $[-1,1]$ (which is true if $\{Q_n'\}$ satisfy a uniform Dini condition). In addition, replace (1.12) by the conditions in the remarks (d) after Definition 1.1. However, one then loses the $O\left(n^{-\tau_1}\right)$ error term, and the asymptotic would hold in compact subsets of $(-1, 1)$.

Finally, we turn to asymptotics for orthonormal polynomials in the plane, and for leading coefficients. We need more notations. Let

$$\phi(z) = z + \sqrt{z^2 - 1} \tag{2.13}$$

denote the conformal map of the exterior of $[-1, 1]$ onto the exterior of the unit ball. For $n \ge 1$, let

$$F_n(\theta) = e^{-2nQ_n(\cos\theta)}|\sin\theta|. \tag{2.14}$$

Define the associated Szegő function

$$D(F_n; z) = \exp\left(\frac{1}{4\pi}\int_{-\pi}^{\pi}\frac{e^{it} + z}{e^{it} - z}\log F_n(t)\,dt\right), |z| < 1. \tag{2.15}$$

**Theorem 2.3.** *Assume that* $\{Q_n\} \in \mathcal{Q}$. *Let* $\varepsilon \in \left(0, \frac{1}{3}\right)$. *There exists* $\tau > 0$ *such that uniformly for* $n \ge 1$ *and* $m$ *satisfying* (2.8),
*(a) For dist* $(z, [-1, 1]) \ge n^{-\tau}$,

$$\left|p_{n,m}(z)\Big/\left\{\frac{1}{\sqrt{2\pi}}\phi(z)^m D^{-1}\left(F_n; \phi(z)^{-1}\right)\right\} - 1\right| \le Cn^{-\tau}. \tag{2.16}$$

*(b) The leading coefficient $\gamma_{n,m}$ of $p_{n,m}$ satisfies*

$$\gamma_{n,m} = \frac{2^m}{\sqrt{\pi}} \exp\left(\frac{n}{\pi} \int_{-1}^{1} Q_n(x) \frac{dx}{\sqrt{1-x^2}}\right)(1 + O(n^{-\tau})). \tag{2.17}$$

*(c) The coefficients $A_{n,m}$ and $B_{n,m}$ in the three term recurrence relation*

$$xp_{n,m}(x) = A_{n,m}p_{n,m-1}(x) + B_{n,m}p_{n,m}(x) + A_{n,m-1}p_{n,m-1}(x)$$

*satisfy*

$$A_{n,m} = \frac{1}{2} + O(n^{-\tau}) \text{ and } B_{n,m} = O(n^{-\tau}). \tag{2.18}$$

*Proof.* (a) See Theorem 13.2(c).
(b) See Theorem 13.1.
(c) See Theorem 13.4.                                                                   □

**Remarks.** (a) Note that for varying weights supported on $[-1, 1]$, Totik's Theorem A is far more general than (2.16) (without the rate).

(b) We expect that one does not need the extra Lipschitz condition (1.12) for Theorem 2.3.

This paper is organized as follows: in Chapter 3, we estimate the equilibrium densities, and some potential theoretic quantities. In Chapter 4, we present restricted range inequalities. In Chapter 5, we provide upper and lower bounds for Christoffel functions and their $L_p$ analogues. In Chapter 6, we obtain upper bounds on the spacing of zeros of orthogonal polynomials, as well as estimates for the largest zeros. In Chapter 7, we establish bounds for orthonormal polynomials. In Chapter 8, we prove a Markov-Bernstein inequality. In Chapters 9 and 10, we use Totik's method to discretize potentials and their derivatives, and then in Chapter 11, we apply these to obtain weighted polynomial approximations. In Chapter 12, we record some identities involving Bernstein-Szegő weights. In Chapter 13, we establish asymptotics for leading coefficients of orthonormal polynomials, as well as uniform asymptotics for orthonormal polynomials and their derivatives, their recurrence coefficients, their zeros, and asymptotics for Christoffel functions. In Chapter 14, we establish further bounds, and in particular lower bounds for the spacing of the zeros. In Chapter 15, we establish universality limits for random matrices associated with $\{e^{-nQ_n}\}$, for fluctuations of eigenvalues, and asymptotics for entropy type integrals.

# Chapter 3
# Potential Theoretic Estimates

Throughout this chapter, we assume that $\{Q_n\} \in \mathcal{Q}$, and use the notation of Definition 1.1. We shall need the Mhaskar-Rakhmanov-Saff numbers for $Q_n$ associated with equilibrium measures of total mass $r$ other than 1. For $r > 0$, define $a_{\pm n,r}$ by

$$
r = \frac{1}{\pi} \int_{a_{-n,r}}^{a_{n,r}} \frac{x Q_n'(x)}{\sqrt{(x - a_{-n,r})(a_{n,r} - x)}} dx;
$$

$$
0 = \frac{1}{\pi} \int_{a_{-n,r}}^{a_{n,r}} \frac{Q_n'(x)}{\sqrt{(x - a_{-n,r})(a_{n,r} - x)}} dx. \tag{3.1}
$$

These are uniquely defined (see Chapter 2 of [25]). As $r$ increases, so does $[a_{-n,r}, a_{n,r}]$, with

$$
\lim_{r \to \infty} a_{n,r} = d_n; \quad \lim_{r \to \infty} a_{-n,r} = c_n.
$$

We also need the equilibrium density on $[a_{-n,r}, a_{n,r}]$ [25, p. 16],

$$
\sigma_{Q_n,r}(x)
$$
$$
= \frac{\sqrt{(x - a_{-n,r})(a_{n,r} - x)}}{\pi^2} \int_{a_{-n,r}}^{a_{n,r}} \frac{Q_n'(s) - Q_n'(x)}{s - x} \frac{ds}{\sqrt{(s - a_{-n,r})(a_{n,r} - s)}}, \tag{3.2}
$$

which has total mass $r$,

$$
\int_{a_{-n,r}}^{a_{n,r}} \sigma_{Q_n,r} = r. \tag{3.3}
$$

© The Author(s) 2018  
E. Levin, D.S. Lubinsky, *Bounds and Asymptotics for Orthogonal Polynomials for Varying Weights*, SpringerBriefs in Mathematics,  
https://doi.org/10.1007/978-3-319-72947-3_3

Its potential

$$V^{\sigma_{Q_{n,r}}}(x) = \int_{a_{-n,r}}^{a_{n,r}} \log \frac{1}{|x-t|} \sigma_{Q_{n,r}}(t)\, dt$$

satisfies

$$V^{\sigma_{Q_{n,r}}} + Q_n \geq F_{Q_{n,r}} \text{ on } I_n; \tag{3.4}$$

$$V^{\sigma_{Q_{n,r}}} + Q_n = F_{Q_{n,r}} \text{ in } [a_{-n,r}, a_{n,r}]. \tag{3.5}$$

In the special case $r = 1$, we have by our hypotheses in Chapter 1,

$$[a_{-n,1}, a_{n,1}] = [-1, 1] \text{ and } \sigma_{Q_{n,1}} = \sigma_{Q_n}. \tag{3.6}$$

We shall need to map $[a_{-n,r}, a_{n,r}]$ to $[-1, 1]$ and scale $\sigma_{Q_{n,r}}$ to a density on $[-1, 1]$. Accordingly, let

$$\beta_{n,r} = \frac{1}{2}(a_{n,r} + a_{-n,r}) \text{ and } \delta_{n,r} = \frac{1}{2}(a_{n,r} - a_{-n,r}) \tag{3.7}$$

and define the map $L_{n,r}$ of $[a_{-n,r}, a_{n,r}]$ onto $[-1, 1]$,

$$L_{n,r}(t) = \frac{t - \beta_{n,r}}{\delta_{n,r}} \text{ with inverse } L_{n,r}^{[-1]}(s) = \beta_{n,r} + \delta_{n,r} s. \tag{3.8}$$

Define the scaled density

$$\hat{\sigma}_{Q_{n,r}}(t) = \frac{\delta_{n,r}}{r} \sigma_{Q_{n,r}} \left( L_{n,r}^{[-1]}(t) \right)$$

$$= \sqrt{1 - t^2} \frac{\delta_{n,r}}{r\pi^2} \int_{-1}^{1} \frac{Q_n' \left( L_{n,r}^{[-1]}(u) \right) - Q_n' \left( L_{n,r}^{[-1]}(t) \right)}{u - t} \frac{du}{\sqrt{1 - u^2}}, \tag{3.9}$$

for $t \in [-1, 1]$. Then

$$\int_{-1}^{1} \hat{\sigma}_{Q_{n,r}}(t)\, dt = 1. \tag{3.10}$$

Also, let

$$U_{n,r}(x) = -\left( V^{\sigma_{Q_{n,r}}}(x) + Q_n(x) - F_{Q_{n,r}} \right), \; x \in I_n. \tag{3.11}$$

We often abbreviate $U_{n,1}$ as $U_n$.

Our main result in this chapter involves estimates for the equilibrium densities:

**Theorem 3.1.** *Let* $\alpha, \alpha_1$ *be as in Definition* 1.1 *and assume* (1.15). *There exists* $r_0 \in \left(\frac{1}{2}, 1\right)$, *such that uniformly for* $r \in \left[r_0, \frac{1}{r_0}\right]$, *and in* $n, t$,
*(a)*

$$\hat{\sigma}_{Q_n,r}(t) \sim \sqrt{1 - t^2}, \quad t \in (-1, 1), \tag{3.12}$$

*and*

$$\sigma_{Q_n,r}(x) \sim \sqrt{(a_{n,r} - x)(x - a_{-n,r})}, \quad x \in [a_{-n,r}, a_{n,r}]. \tag{3.13}$$

*(b)*

$$\left| \hat{\sigma}_{Q_n,r}(t) / \sqrt{1 - t^2} - \hat{\sigma}_{Q_n,r}(s) / \sqrt{1 - s^2} \right| \leq C |s - t|^\alpha, \quad s, t \in (-1, 1), \tag{3.14}$$

*and*

$$\left| \hat{\sigma}_{Q_n,r}(t) - \hat{\sigma}_{Q_n,r}(s) \right| \leq C |s - t|^\alpha, \quad s, t \in (-1, 1). \tag{3.15}$$

**Remarks.** (a) Applying Fatou's lemma in (3.9) gives

$$\liminf_{t \to 1-} \hat{\sigma}_{Q_n,r}(t) / \sqrt{1 - t^2}$$

$$\geq \frac{\delta_{n,r}}{r\pi^2} \int_{-1}^{1} \frac{Q_n'(a_{n,r}) - Q_n'(a_{n,r} - \delta_{n,r}(1 - u))}{(1 - u)^{3/2}} \frac{du}{\sqrt{1 + u}}.$$

Here we also are using that the integrand in (3.9) is nonnegative as $Q_n'$ is increasing. In particular,

$$\liminf_{t \to 1-} \sigma_{Q_n,1}(t) / \sqrt{1 - t^2} \geq \frac{1}{\pi^2} \int_{-1}^{1} \frac{Q_n'(1) - Q_n'(u)}{(1 - u)^{3/2}} \frac{du}{\sqrt{1 + u}}.$$

It then follows from the bound $\sigma_{Q_n,1}(t) \leq C\sqrt{1 - t^2}, t \in (-1, 1)$, that as $x \to 1-$,

$$\frac{Q_n'(1) - Q_n'(x)}{(1 - x)^{1/2}} \leq 2^{3/2} \int_{x-(1-x)}^{x} \frac{Q_n'(1) - Q_n'(u)}{(1 - u)^{3/2}} du \to 0.$$

Thus

$$Q_n'(1) - Q_n'(x) = o(1 - x)^{1/2},$$

as $x \to 1-$. This explains our more severe smoothness assumption on $Q_n'$ near $\pm 1$.

(b) Define the Hilbert transform on the real line, for integrable functions $f : \mathbb{R} \to \mathbb{R}$ by

$$H[f](x) = \frac{1}{\pi} PV \int_{-\infty}^{\infty} \frac{f(t)}{t-x} dt,$$

a.e. $x \in \mathbb{R}$. It is a classic result [50, p. 122] that

$$H[H[f]] = -f,$$

at least for square integrable functions. We can recast (3.2) for $r = 1$, as

$$\frac{\pi \sigma_{Q_n}(x) \chi(x)}{\sqrt{1-x^2}} = H\left[ \frac{Q_n'(\cdot)}{\sqrt{1-\cdot^2}} \chi \right](x),$$

where $\chi$ is the characteristic function of $[-1, 1]$. Then the invertibility of $H$ gives

$$\frac{Q_n'(x)}{\sqrt{1-x^2}} \chi(x) = -\pi H\left[ \frac{\pi \sigma_{Q_n}(\cdot) \chi}{\sqrt{1-\cdot^2}} \right](x).$$

It follows from Privalov's Theorem [60, Vol. I, p. 121, Thm. 13.29] that if $\sigma_{Q_n}$ satisfies a Lipschitz condition of some positive order in each compact subset of $(-1, 1)$, then the same is true of $\frac{Q_n'(x)}{\sqrt{1-x^2}}$. Moreover, if this is uniform in $n$ for $\sigma_{Q_n}$, it will be uniform in $n$ for $Q_n'$.

We begin by proving estimates for $a_{\pm n,r}$ and $\hat{\sigma}_{Q_n,r}$:

**Lemma 3.2.** *(a) For $n \geq 1$, $r > 0$,*

$$Q_n'(a_{n,r}) \delta_{n,r} \geq r \text{ and } Q_n'(a_{-n,r}) \delta_{n,r} \leq -r. \tag{3.16}$$

*Furthermore, there exists $C > 0$ such that*

$$C \geq Q_n'(1) \geq 1 \text{ and } -C \leq Q_n'(-1) \leq -1. \tag{3.17}$$

*(b) Uniformly in $r \in \left[ \frac{1}{2}, 2 \right]$ and in $n$,*

$$\delta_{n,r} \sim 1, \tag{3.18}$$

*Proof.* (a) We can reformulate (3.1) as

$$r = \frac{1}{\pi} \int_{a_{-n,r}}^{a_{n,r}} Q_n'(x) \sqrt{\frac{x - a_{-n,r}}{a_{n,r} - x}} dx; \tag{3.19}$$

$$-r = \frac{1}{\pi} \int_{a_{-n,r}}^{a_{n,r}} Q_n'(x) \sqrt{\frac{a_{n,r} - x}{x - a_{-n,r}}} dx. \tag{3.20}$$

Since for $a < b$,

$$\frac{1}{\pi} \int_a^b \sqrt{\frac{x-a}{b-x}} dx = \frac{1}{2}(b-a) = \frac{1}{\pi} \int_a^b \sqrt{\frac{b-x}{x-a}} dx,$$

and since $Q_n'$ is increasing, this also gives

$$r \leq Q_n'(a_{n,r}) \delta_{n,r} \text{ and } -r \geq Q_n'(a_{-n,r}) \delta_{n,r}.$$

Next, the uniform Lipschitz condition on $Q_n'$ gives

$$\left| Q_n'(1) \right| + \left| Q_n'(-1) \right| = \left| Q_n'(1) - Q_n'(-1) \right| \leq C(2)^\alpha,$$

and then since $\delta_{n,1} = 1$, we also obtain (3.17).

(b) First let $t_n$ denote a point with

$$Q_n'(t_n) = 0. \tag{3.21}$$

This exists, since $Q_n$ is convex and $Q_n$ has limit $\infty$ at $c_n+$ and $d_n-$. Note that $Q_n'(t)(t-t_n) \geq 0$ in $I_n$. From (3.16), we see that for $r > 0$,

$$t_n \in (a_{-n,r}, a_{n,r}). \tag{3.22}$$

It then follows from [25, Thm. 2.4(iii), p. 41] that

$$t_n = a_{n,0} := \lim_{r \to 0} a_{n,r}.$$

Note that there $a_{n,0} = 0$, but we can translate the interval to our case. For future use, we observe that

$$1 \leq Q_n'(1) = Q_n'(1) - Q_n'(t_n) \leq C(1 - t_n)^\alpha$$

and a similar inequality holds for $1 + t_n$. Thus for some $0 < C < 1$,

$$t_n \in [-1 + C, 1 - C]. \tag{3.23}$$

Then (3.22) shows that

$$a_{n,r} \in [-1 + C, 1]; \ a_{-n,r} \in [-1, 1 - C] \text{ for } r \in (0, 1). \tag{3.24}$$

We next establish an upper bound for $\delta_{n,r}$. If $r \leq 1$, then $\delta_{n,r} \leq \delta_{n,1} = 1$ (cf. [25, Thm. 2.4(iii), p. 41]). Suppose now $r \in (1, 2]$. If $\max\{a_{n,r}, |a_{-n,r}|\} \leq 2$, we are done. Suppose this fails and say $a_{n,r} \geq |a_{-n,r}|$ and $a_{n,r} > 2$. Now (3.1) and our assumption that $a_{n,r} > 2$ give

$$r = \frac{1}{\pi} \int_{a_{-n,r}}^{a_{n,r}} \frac{(x - t_n)\, Q'_n(x)}{\sqrt{(x - a_{-n,r})(a_{n,r} - x)}}\, dx$$

$$\geq \frac{1}{\pi} \int_{\frac{2}{3}a_{n,r}}^{a_{n,r}} \frac{\left(\frac{2}{3}a_{n,r} - 1\right) Q'_n(1)}{\sqrt{2\delta_{n,r}(a_{n,r} - x)}}\, dx$$

$$\geq C a_{n,r} \int_{\frac{2}{3}a_{n,r}}^{a_{n,r}} \frac{dx}{\sqrt{2\delta_{n,r}(a_{n,r} - x)}} \geq C\frac{a_{n,r}^{3/2}}{\sqrt{\delta_{n,r}}} \geq C a_{n,r}.$$

Then still $\delta_{n,r} \leq a_{n,r} \leq C$.

Next, we establish a lower bound for $\delta_{n,r}$. If $r \in [1, 2]$, we have $\delta_{n,r} \geq \delta_{n,1} = 1$. Suppose next that $r \in \left[\frac{1}{2}, 1\right]$. As $Q'_n$ is increasing, (a) gives

$$\frac{1}{2} \leq r \leq Q'_n(a_{n,r})\, \delta_{n,r} \leq Q'_n(1)\, \delta_{n,r} \leq C\delta_{n,r}.$$

$\square$

**Lemma 3.3.** *(a) For* $0 < r < s$,

$$\frac{\delta_{n,s}}{s} \leq \frac{\delta_{n,r}}{r}, \tag{3.25}$$

*and*

$$|a_{n,s} - a_{n,r}| + |a_{-n,s} - a_{-n,r}| \leq 2\frac{\delta_{n,s}}{s}(s - r). \tag{3.26}$$

*(b) There exists* $r_1 \in (0, 1)$ *such that for* $n \geq n_1$ *and* $r, s \in \left[r_1, \frac{1}{r_1}\right]$,

$$|a_{n,s} - a_{n,r}| \sim s - r \text{ and } |a_{-n,s} - a_{-n,r}| \sim s - r. \tag{3.27}$$

*Proof.* (a) First note that the strict monotonicity of $a_{n,r}$ in $r$ is proved in Chapter 2 of [25, Thm. 2.4(iii), p. 41]. Fix $0 < r < s$ and define the linear map

$$\mathcal{L}(x) = L_{n,r}^{[-1]} \circ L_{n,s}(x).$$

It maps $[a_{-n,s}, a_{n,s}]$ onto $[a_{-n,r}, a_{n,r}]$. In particular,

$$\mathcal{L}(a_{-n,s}) = a_{-n,r} > a_{-n,s};$$
$$\mathcal{L}(a_{n,s}) = a_{n,r} < a_{n,s}.$$

Then the equation $\mathcal{L}(x) = x$ has a unique root $\tau_n \in (a_{-n,s}, a_{n,s})$. By our equilibrium relations (3.1), and then a substitution $x = \mathcal{L}(y)$,

$$
\begin{aligned}
r &= \frac{1}{\pi} \int_{a_{-n,r}}^{a_{n,r}} \frac{(x - \mathcal{L}(\tau_n)) Q_n'(x)}{\sqrt{(x - a_{-n,r})(a_{n,r} - x)}} dx \\
&= \frac{1}{\pi} \int_{a_{-n,s}}^{a_{n,s}} \frac{(\mathcal{L}(y) - \mathcal{L}(\tau_n)) Q_n'(\mathcal{L}(y))}{\sqrt{(y - a_{-n,s})(a_{n,s} - y)}} dy \\
&= \frac{\delta_{n,r}}{\delta_{n,s}} \frac{1}{\pi} \int_{a_{-n,s}}^{a_{n,s}} \frac{(y - \tau_n) Q_n'(\mathcal{L}(y))}{\sqrt{(y - a_{-n,s})(a_{n,s} - y)}} dy.
\end{aligned}
\tag{3.28}
$$

Now for $y < \tau_n$, we have $y - \tau_n < 0$ and $Q_n'(\mathcal{L}(y)) > Q_n'(y)$, since $\mathcal{L}(y) > y$, so

$$
(y - \tau_n) Q_n'(\mathcal{L}(y)) < (y - \tau_n) Q_n'(y).
\tag{3.29}
$$

Similarly for $y > \tau_n$, we have $y - \tau_n > 0$ and $Q_n'(\mathcal{L}(y)) < Q_n'(y)$, so again (3.29) holds. Then (3.28) gives

$$
r \le \frac{\delta_{n,r}}{\delta_{n,s}} \frac{1}{\pi} \int_{a_{-n,s}}^{a_{n,s}} \frac{(y - \tau_n) Q_n'(y)}{\sqrt{(y - a_{-n,s})(a_{n,s} - y)}} dy = \frac{\delta_{n,r}}{\delta_{n,s}} s,
$$

by our equilibrium relations (3.1). So we have (3.25). Finally, we also then obtain

$$
\delta_{n,s} - \delta_{n,r} \le \delta_{n,s} \left(1 - \frac{r}{s}\right),
$$

which easily yields (3.26).

(b) We use the identity [25, p. 62, eqn. (2.75)]

$$
\frac{1}{2} \frac{da_{n,t}}{dt} \left\{ Q_n'(a_{n,t}) + \frac{1}{\pi} \int_{a_{-n,t}}^{a_{n,t}} \frac{Q_n'(a_{n,t}) - Q_n'(x)}{(a_{n,t} - x)^{3/2}} (x - a_{-n,t})^{1/2} dx \right\} = 1.
$$

This is valid provided $a_{n,t}$ is in the neighborhood of 1 in which $Q_n'$ satisfies the Lipschitz condition of order $\alpha_1 > \frac{1}{2}$. By (a), there exists $r_1 \in (0, 1)$ such that for $t \in \left[r_1, \frac{1}{r_1}\right]$, we have this inclusion. Note that in [25], it was assumed that $Q'(0) = 0$, but we can apply the result there to $Q_n(t - t_n)$. Straightforward estimation, using our Lipschitz condition, using (3.18) and the locally uniform boundedness of $\{Q_n'\}$ gives

$$
C_1 \le \frac{da_{n,t}}{dt} \le C_2, \quad n \ge 1, t \in \left[r_1, \frac{1}{r_1}\right].
$$

Then $|a_{n,s} - a_{n,r}| \sim |r - s|$ and the analogous inequality on the left is similar. $\square$

We need one more lemma, a Privalov type estimate. Presumably the second part is in the literature somewhere, but we have been unable to find it.

**Lemma 3.4.** *Let $h : [-1, 1] \to \mathbb{R}$ satisfy a Lipschitz condition of order $\alpha \in (0, 1)$ on $[-1, 1]$. Assume, moreover, that in a neighborhood of $\pm 1$, $h$ satisfies a Lipschitz condition of order $\alpha_1 > \frac{1}{2}$. Define*

$$H(x) = PV \int_{-1}^{1} \frac{h(t)}{(t-x)\sqrt{1-t^2}} dt, \quad x \in (-1, 1),$$

*and define $H$ by its limiting values at $\pm 1$. Then $H$ satisfies a Lipschitz condition of order $\alpha$ in each compact subset of $(-1, 1)$, and for some $\varepsilon \in \left(0, \frac{1}{4}\right)$, satisfies a Lipschitz condition of order $\alpha_1 - \frac{1}{2}$ in $[-1, -1+\varepsilon] \cup [1-\varepsilon, 1]$.*

*Proof.* Since $t \to h(t)/\sqrt{1-t^2}$ satisfies a Lipschitz condition of order $\alpha$ in each compact subset of $(-1, 1)$ (recall (1.15)), the first assertion follows from the classic Privalov theorem for principal value integrals on arcs [60, Vol. I, p. 121, Thms. 13.29, 13.30]. We turn to the second part, and prove it on $[1-\varepsilon, 1]$. Since the function

$$\int_{-1}^{1-2\varepsilon} \frac{h(t)}{(t-x)\sqrt{1-t^2}} dt$$

is a continuously differentiable function of $x \in [1-\varepsilon, 1]$, we can simply assume that $h$ satisfies a Lipschitz condition of order $\alpha_1 > \frac{1}{2}$ throughout $[-1, 1]$. Define

$$G(t) = \frac{h(t) - h(1)}{\sqrt{1-t^2}}, t \in (-1, 1).$$

Observe that $G$ has limit 0 as $t \to 1-$. Define $G = 0$ in $[1, 2]$. Then for $x \in (-1, 2)$,

$$H(x) = PV \int_{-1}^{2} \frac{G(t)}{t-x} dt.$$

If we can show that $G$ satisfies a Lipschitz condition of order $\alpha_1 - \frac{1}{2}$ in $[0, 2]$, then the classic Privalov theorem for an arc that we quoted above shows that $H$ satisfies the requisite Lipschitz condition of order $\alpha_1 - \frac{1}{2}$ in $[1-\varepsilon, 1]$. Obviously we need to only show that $G$ satisfies the Lipschitz condition in $[0, 1]$. So let $0 \le s < t \le 1$. We consider two subcases:
(I) $1 - t \ge \frac{1}{2}(1-s)$. We see that

$$|G(s) - G(t)| = \left| \frac{h(s) - h(t)}{\sqrt{1-s^2}} + [h(t) - h(1)] \left[ \frac{1}{\sqrt{1-s^2}} - \frac{1}{\sqrt{1-t^2}} \right] \right|$$

$$\le C \frac{|s-t|^{\alpha_1}}{\sqrt{1-s}} + C|t-1|^{\alpha_1} \frac{|s-t|}{(1-s)\sqrt{1-t}}.$$

Now $t - s \leq 1 - s$ and $1 - t \leq 1 - s$. Then (as $\alpha_1 < 1$),

$$
\begin{aligned}
|G(s) - G(t)| &\leq C |s - t|^{\alpha_1 - \frac{1}{2}} + C |t - 1|^{\alpha_1 - 3/2} |s - t| \\
&\leq C |s - t|^{\alpha_1 - \frac{1}{2}} + C |s - 1|^{\alpha_1 - 3/2} |s - t| \\
&\leq C |s - t|^{\alpha_1 - \frac{1}{2}} .
\end{aligned}
$$

(II) $1 - t \leq \frac{1}{2} (1 - s)$. Then

$$
|s - t| = |(1 - t) - (1 - s)| \geq \frac{1}{2} (1 - s) \geq 1 - t.
$$

Then

$$
\begin{aligned}
|G(s) - G(t)| &\leq |G(s)| + |G(t)| \\
&\leq C |s - 1|^{\alpha_1 - \frac{1}{2}} + C |t - 1|^{\alpha_1 - \frac{1}{2}} \\
&\leq C |s - t|^{\alpha_1 - \frac{1}{2}} .
\end{aligned}
$$

$\square$

For the proof of Theorem 3.1, we apply Lemma 3.3 with $s = 1$, so that $a_{\pm n, s} = \pm 1$ to deduce that for $r > 0$,

$$
|a_{n,r} - 1| + |a_{-n,r} + 1| \leq 2 |r - 1| . \tag{3.30}
$$

Then we can choose $r_0 \in \left( \frac{1}{2}, 1 \right)$ and $\eta_0 \in \left( 0, \frac{1}{4} \right)$ such that if $I_0$ is as in Definition 1.1, then for $r \in \left[ r_0, \frac{1}{r_0} \right]$ and $n \geq 1$, then

$$
L_{n,r}^{[-1]} [1 - 2\eta_0, 1] \subset I_0 \cap I_n . \tag{3.31}
$$

*Proof of Theorem* 3.1. (a) Suppose first that $|t| \leq 1 - \eta_0$ with $\eta_0$ as above. The uniform Lipschitz condition (1.11), (3.9), and (3.18) give

$$
\begin{aligned}
\hat{\sigma}_{Q_n,r}(t) / \sqrt{1 - t^2} &\leq C \delta_{n,r} \int_{-1}^{1} \delta_{n,r}^{\alpha} |u - t|^{\alpha - 1} \frac{du}{\sqrt{1 - u^2}} \\
&\leq \frac{C}{\sqrt{\eta_0}} \int_{-1 + \eta_0/2}^{1 - \eta_0/2} |u - t|^{\alpha - 1} du + C \eta_0^{\alpha - 1} \int_{1 - \eta_0/2 \leq |u| \leq 1} \frac{du}{\sqrt{1 - u^2}} \\
&\leq \frac{C}{\sqrt{\eta_0}} \int_{-2}^{2} |s|^{\alpha - 1} ds + C \eta_0^{\alpha - 1} \int_{1 - \eta_0/2 \leq |u| \leq 1} \frac{du}{\sqrt{1 - u^2}} \leq C_2 .
\end{aligned}
$$

Let us next assume $t \in [1 - \eta_0, 1]$. Using (3.9), (1.12) and the substitution $1 - u = (1 - t) s$, we estimate

$$
\hat{\sigma}_{Q_n,r}(t) / \sqrt{1 - t^2} \leq C \eta_0^{\alpha-1} \int_{|u| \leq 1 - 2\eta_0} \frac{du}{\sqrt{1 - u^2}} + C \int_{1-2\eta_0}^1 |u - t|^{\alpha_1 - 1} \frac{du}{\sqrt{1 - u}}
$$

$$
\leq C + C (1 - t)^{\alpha_1 - 1/2} \int_0^{2\eta_0/(1-t)} |1 - s|^{\alpha_1 - 1} \frac{ds}{\sqrt{s}}
$$

$$
\leq C + C (1 - t)^{\alpha_1 - 1/2} \left\{ \int_0^{\min\{2, 2\eta_0/(1-t)\}} |1 - s|^{\alpha_1 - 1} \frac{ds}{\sqrt{s}} \right.
$$
$$
\left. + \int_{\min\{2, 2\eta_0/(1-t)\}}^{2\eta_0/(1-t)} s^{\alpha_1 - \frac{3}{2}} ds \right\}
$$

$$
\leq C + C (1 - t)^{\alpha_1 - 1/2} \left\{ 1 + (1 - t)^{-(\alpha_1 - \frac{1}{2})} \right\} \leq C,
$$

as $\alpha_1 > \frac{1}{2}$. So we have the upper bound implicit in (3.12). We turn to the matching lower bound. Suppose that $t$ is given with $Q_n'\left(L_{n,r}^{[-1]}(t)\right) \geq 0$, so $L_{n,r}^{[-1]}(t) \geq t_n$. We have $Q_n'\left(L_{n,r}^{[-1]}(u)\right) \leq 0$ for $u \leq L_{n,r}(t_n)$, so

$$
\hat{\sigma}_{Q_n,r}(t) / \sqrt{1 - t^2} \geq \frac{\delta_{n,r}}{r \pi^2} \int_{-1}^{L_{n,r}(t_n)} \frac{Q_n'\left(L_{n,r}^{[-1]}(u)\right) - Q_n'\left(L_{n,r}^{[-1]}(t)\right)}{u - t} \frac{du}{\sqrt{1 - u^2}}
$$

$$
\geq \frac{\delta_{n,r}}{r \pi^2} \int_{-1}^{L_{n,r}(t_n)} \frac{\left|Q_n'\left(L_{n,r}^{[-1]}(u)\right)\right| + Q_n'\left(L_{n,r}^{[-1]}(t)\right)}{2} du
$$

$$
\geq C \delta_{n,r} \int_{-1}^{L_{n,r}(t_n)} \left|Q_n'\left(L_{n,r}^{[-1]}(u)\right)\right| du
$$

$$
= C \int_{a_{-n,r}}^{t_n} \left|Q_n'(s)\right| ds.
$$

Now (3.16) shows that $|Q'(a_{-n,r})| \geq C$. Then our uniform Lipschitz condition shows that $Q_n'(s) < 0$ and $\left|Q_n'(s)\right| \geq \frac{C}{2}$ for $s \in [a_{-n,r}, a_{-n,r} + \eta_2]$, where $\eta_2$ is independent of $r, n$. Then

$$
\int_{a_{-n,r}}^{t_n} \left|Q'(s)\right| ds \geq C_1
$$

independently of $n, r$. The case where $Q_n'\left(L_{n,r}^{[-1]}(t)\right) < 0$ is similar. So we have proved (3.12). Then (3.13) follows using (3.18).

(b) We apply Lemma 3.4 to $\hat{\sigma}_{Q_n,r}(x) / \sqrt{1 - x^2}$, in the form given by (3.9). Recall too that $\alpha < \alpha_1 - \frac{1}{2}$. Then (3.14) follows. Since $t \to \sqrt{1 - t^2}$ satisfies a Lipschitz condition of order $\frac{1}{2}$, we also obtain (3.15).                               $\square$

Next we turn to growth of potentials:

**Lemma 3.5.** *The potential $V^{\hat{\sigma}_{Q_n,r}}$ satisfies for $n \geq 1$, $r \in \left[r_0, \frac{1}{r_0}\right]$, $x \in \mathbb{R}$, $|y| \leq 1$,*

$$0 \geq V^{\hat{\sigma}_{Q_n,r}}(x+iy) - V^{\hat{\sigma}_{Q_n,r}}(x)$$

$$\geq -C|y| \left\{ \sqrt{\max\{1-|x|,0\}} + \sqrt{|y|} \right\}. \tag{3.32}$$

*The constant $C$ is independent of $n, r, x, y$.*

*Proof.* Let us first assume that $x \in [0, 1]$. Then using (3.12),

$$V^{\hat{\sigma}_{Q_n,r}}(x+iy) - V^{\hat{\sigma}_{Q_n,r}}(x) = -\frac{1}{2} \int_{-1}^{1} \log\left[1 + \left(\frac{y}{x-s}\right)^2\right] \hat{\sigma}_{Q_n,r}(s)\, ds$$

$$\geq -C \int_{0}^{1} \log\left[1 + \left(\frac{y}{x-s}\right)^2\right] \sqrt{1-s}\, ds$$

$$\geq -C \int_{0}^{1} \log\left[1 + \left(\frac{y}{x-s}\right)^2\right] \left(\sqrt{1-x} + \sqrt{|x-s|}\right) ds$$

$$\geq -C|y| \int_{-\infty}^{\infty} \log\left[1 + \left(\frac{1}{u}\right)^2\right] \left(\sqrt{1-x} + \sqrt{|y|}\sqrt{|u|}\right) du,$$

by the substitution $s - x = u|y|$. This gives (3.32) for $x \in [0, 1]$. Next, if $x \in (1, \infty)$,

$$V^{\hat{\sigma}_{Q_n,r}}(x+iy) - V^{\hat{\sigma}_{Q_n,r}}(x) = -\frac{1}{2} \int_{-1}^{1} \log\left[1 + \left(\frac{y}{x-s}\right)^2\right] \hat{\sigma}_{Q_n,r}(s)\, ds$$

$$\geq -C \int_{0}^{1} \log\left[1 + \left(\frac{y}{1-s}\right)^2\right] \sqrt{1-s}\, ds$$

$$= -C|y|^{3/2} \int_{0}^{1/y} \log\left[1 + \frac{1}{t^2}\right] \sqrt{t}\, dt,$$

by the substitution $1 - s = t|y|$. So we also have the result for $x \in [1, \infty)$. The case $x < 0$ is similar.  □

Next, we study the behavior of $U_{n,r}$ close to $a_{n,r}$. Recall from (3.11) that

$$U_{n,r}(x) = -V^{\sigma_{Q_n,r}}(x) - Q_n(x) + F_{Q_n,r}, \quad x \in I_n,$$

and that $U_{n,r} \leq 0$ in $I_n$, with equality in $[a_{-n,r}, a_{n,r}]$. We shall need Green's functions. Recall that given an interval $\Delta$, and $x \notin \Delta$, $g_\Delta(z, x)$ denotes the Green's function for $\overline{\mathbb{C}} \backslash \Delta$ with pole at $x$, so that $g_\Delta(z, x) + \log|z - x|$ is harmonic as a function of $z$ in $\overline{\mathbb{C}}\backslash\Delta$ and vanishes on $\Delta$. When $x \in \Delta$, we set $g_\Delta(z,x) \equiv 0$, and when $x = \infty$, the Green's function is denoted by $g_\Delta(z)$. For example, if $\Delta = [-1, 1]$, then $g_{[-1,1]}(z) = \log\left|z + \sqrt{z^2 - 1}\right|$.

**Lemma 3.6.** *(a) There exists $\eta_2 > 0$, such that for $n \geq 1$, $r \in \left[r_0, \frac{1}{r_0}\right]$, and $t \in [1, 1 + \eta_2] \cap I_0$,*

$$U_{n,r}\left(L_{n,r}^{[-1]}(t)\right) \sim -(t-1)^{3/2}. \tag{3.33}$$

*Moreover for $n \geq 1$, $r \in \left[r_0, \frac{1}{r_0}\right]$, and $x \in [a_{n,r}, a_{n,r} + \eta_2] \cap I_0$,*

$$U_{n,r}(x) \sim -(x - a_{n,r})^{3/2}. \tag{3.34}$$

*(b) Let $B > 0$, $\rho \in (a_{n,r}, a_{n,r} + \eta_2]$. For $x \in [\rho, d_n)$, and some $C_1, C_2$ independent of $n, r, B$,*

$$nU_{n,r}(x) + Bg_{[a_{-n,r},a_{n,r}]}(x) \leq -C_1 n (\rho - a_{n,r})^{3/2} + BC_2 (\rho - a_{n,r})^{1/2}, \tag{3.35}$$

*and for $n \geq C_6 B/(\rho - a_{n,r})$,*

$$nU_{n,r}(x) + Bg_{[a_{-n,r},a_{n,r}]}(x) \leq -C_7 n (\rho - a_{n,r})^{3/2}. \tag{3.36}$$

*(c) In particular, for some $C_7$ independent of $n, r \in \left[r_0, \frac{1}{r_0}\right], B$,*

$$\sup_{x \in I_n}\left(nU_{n,r}(x) + Bg_{[a_{-n,r},a_{n,r}]}(x)\right) \leq C_7 Bn^{-1/2}. \tag{3.37}$$

*Similar inequalities hold in the interval $(c_n, -\rho)$.*

*Proof.* (a) We know that

$$-\int_{a_{-n,r}}^{a_{n,r}} \log|x - t|\, \sigma_{Q_n,r}(t)\, dt + Q_n(x) = F_{Q_n,r}, x \in [a_{-n,r}, a_{n,r}].$$

Since $\sigma_{Q_n,r}$ satisfies a Lipschitz condition of positive order in $[-1, 1]$, we can differentiate this relation to obtain (cf. [32, p. 29 ff.])

$$-PV \int_{a_{-n,r}}^{a_{n,r}} \frac{1}{x - t} \sigma_{Q_n,r}(t)\, dt + Q_n'(x) = 0, \quad x \in (a_{-n,r}, a_{n,r}).$$

Here $PV$ stands for Cauchy principal value. Since $\{\sigma_{Q_n,r}\}$ satisfy a uniform in $n, r$ Lipschitz condition in $(a_{-n,r}, a_{n,r})$, and since $\hat{\sigma}_{Q_n,r}(t) \sim \sqrt{1 - t^2} \Rightarrow \sigma_{Q_n,r}(x) \sim \sqrt{(x - a_{-n,r})(a_{n,r} - x)}$ uniformly in $n, r$ and $x$, we can let $x \to a_{n,r}-$ to obtain

$$-\int_{a_{-n,r}}^{a_{n,r}} \frac{1}{a_{n,r} - t} \sigma_{Q_n,r}(t)\, dt + Q_n'(a_{n,r}) = 0.$$

(The justification is straightforward and left to the reader.) We thus have both $U_{n,r}(a_{n,r}) = 0$ and $U_{n,r}'(a_{n,r}) = 0$. Also for $x > a_{n,r}$,

$$U_{n,r}'(x) = \int_{a_{-n,r}}^{a_{n,r}} \frac{1}{x - t} \sigma_{Q_n,r}(t)\, dt - Q_n'(x)$$

is a decreasing function of $x$, and in particular, $U_{n,r}'(x) < 0$. Then if $x \in (a_{n,r}, d_n) \cap I_0$, we have for some $\xi \in (a_{n,r}, x)$,

$$U_{n,r}(x) = U_{n,r}(a_{n,r}) + (x - a_{n,r}) U_{n,r}'(\xi)$$
$$\geq 0 + (x - a_{n,r}) U_{n,r}'(x). \tag{3.38}$$

Here

$$U_{n,r}'(x) = U_{n,r}'(x) - U_{n,r}'(a_{n,r})$$
$$= \int_{a_{-n,r}}^{a_{n,r}} \left[ \frac{1}{x - t} - \frac{1}{a_{n,r} - t} \right] \sigma_{Q_n,r}(t)\, dt - \left[ Q_n'(x) - Q_n'(a_{n,r}) \right].$$

Using our Lipschitz condition (1.12), and (3.13) and that $x \in I_0$, we continue this as

$$= -(x - a_{n,r}) \int_{a_{-n,r}}^{a_{n,r}} \frac{1}{(x - t)(a_{n,r} - t)} \sigma_{Q_n,r}(t)\, dt - \left[ Q_n'(x) - Q_n'(a_{n,r}) \right]$$
$$\geq -C(x - a_{n,r}) \left[ \int_0^{a_{n,r}} \frac{1}{(x - t)(a_{n,r} - t)} \sqrt{(t - a_{-n,r})(a_{n,r} - t)}\, dt + 1 \right]$$
$$\quad - C_2 (x - a_{n,r})^{\alpha_1}$$
$$\geq -C(x - a_{n,r}) \left[ \int_0^{a_{n,r}} \frac{1}{(x - t)\sqrt{a_{n,r} - t}} \sqrt{\delta_{n,r}}\, dt + 1 \right] - C_2 (x - a_{n,r})^{\alpha_1}$$
$$= -C(x - a_{n,r})^{1/2} \int_0^{\frac{a_{n,r}}{x - a_{n,r}}} \frac{1}{(1 + s)\sqrt{s}}\, ds - C_2 (x - a_{n,r})^{\alpha_1}$$
$$\geq -C_1 (x - a_{n,r})^{1/2} - C_2 (x - a_{n,r})^{\alpha_1}.$$

In the second last line, we made the substitution $(a_{n,r} - t) = (x - a_{n,r}) s$. The constants don't depend on $n$ or $x$. As $\alpha_1 > \frac{1}{2}$, we have for some $C_3, \eta_2 > 0$ independent of $n, r$, we have

$$U'_{n,r}(x) \geq -C_3 (x - a_{n,r})^{1/2}, \quad x \in [a_{n,r}, a_{n,r} + \eta_2] \cap I_0.$$

Then from (3.38),

$$U_{n,r}(x) \geq -C_4 (x - a_{n,r})^{3/2}, \quad x \in [a_{n,r}, a_{n,r} + \eta_2] \cap I_0.$$

Similarly, for some $\zeta$ between $x (\geq a_{n,r})$ and $\frac{a_{n,r} + x}{2}$

$$U_{n,r}(x) = U_{n,r}\left(\frac{a_{n,r} + x}{2}\right) + \left(\frac{x - a_{n,r}}{2}\right) U'_{n,r}(\zeta)$$

$$\leq 0 + \left(\frac{x - a_{n,r}}{2}\right) U'_{n,r}\left(\frac{a_{n,r} + x}{2}\right)$$

$$= \left(\frac{x - a_{n,r}}{2}\right) \left\{ \int_{a_{-n,r}}^{a_{n,r}} \left[ \frac{1}{\frac{a_{n,r} + x}{2} - t} - \frac{1}{a_{n,r} - t} \right] \sigma_{Q_{n,r}}(t)\, dt \right.$$

$$\left. - \left[ Q'_n\left(\frac{a_{n,r} + x}{2}\right) - Q'_n(a_{n,r}) \right] \right\}$$

$$\leq \left(\frac{x - a_{n,r}}{2}\right) \int_{a_{-n,r}}^{a_{n,r}} \left[ \frac{1}{\frac{a_{n,r} + x}{2} - t} - \frac{1}{a_{n,r} - t} \right] \sigma_{Q_{n,r}}(t)\, dt$$

$$\leq -C (x - a_{n,r})^{3/2},$$

as above. So we have (3.34). Since

$$L_{n,r}^{[-1]}(t) - a_{n,r} = \delta_{n,r}(t - 1),$$

we also obtain (3.33).

(b) Now for $x > a_{n,r}$,

$$\frac{d}{dx}\left[ n U_{n,r}(x) + B g_{[a_{-n,r}, a_{n,r}]}(x) \right]$$

$$= n U'_{n,r}(x) + B \int_{a_{-n,r}}^{a_{n,r}} \frac{1}{x - t} \frac{dt}{\pi \sqrt{(t - a_{-n,r})(a_{n,r} - t)}}$$

is a decreasing function of $x$, so for $x \in [\rho, d_n)$, standard estimates for $g_{[a_{-n,r}, a_{n,r}]}$ give,

$$nU_{n,r}(x) + Bg_{[a_{-n,r}, a_{n,r}]}(x) \leq nU_n(\rho) + Bg_{[a_{-n,r}, a_{n,r}]}(\rho)$$
$$\leq -C_1 n (\rho - a_{n,r})^{3/2} + BC_2 (\rho - a_{n,r})^{1/2}$$
$$= (\rho - a_{n,r})^{1/2} [-C_1 n (\rho - a_{n,r}) + BC_2].$$

Then (3.35) and (3.36) follow.

(c) From (b), with $\rho = a_{n,r} + \frac{1}{n}$, we have for $x \in [a_{n,r} + \frac{1}{n}, d_n)$,

$$nU_{n,r}(x) + Bg_{[a_{-n,r}, a_{n,r}]}(x) \leq (-C_1 + BC_2) n^{-1/2}.$$

Also for such $n$ and $x \in [a_{n,r}, a_{n,r} + \frac{1}{n}]$, the estimate $g_{[a_{-n,r}, a_{n,r}]}(x) \leq C\sqrt{x - a_{n,r}}$ gives

$$nU_{n,r}(x) + Bg_{[a_{-n,r}, a_{n,r}]}(x) \leq 0 + BC_2 n^{-1/2}.$$

Thus

$$\sup_{x \in [a_{n,r}, d_n)} \left( nU_n(x) + Bg_{[a_{-n,r}, a_{n,r}]}(x) \right) \leq C_7 Bn^{-1/2}.$$

A similar estimate holds in $(c_n, a_{-n,r}]$. Since in $[a_{-n,r}, a_{n,r}]$, we have $nU_{n,r}(x) + Bg_{[a_{-n,r}, a_{n,r}]}(x) = 0$, the result follows.  $\square$

# Chapter 4
# Restricted Range Inequalities

In the theory of weighted polynomials, the Mhaskar-Saff identity

$$\left\| Pe^{-nQ_n} \right\|_{L_\infty(I)} = \left\| Pe^{-nQ_n} \right\|_{L_\infty[-1,1]},$$

valid for polynomials $P$ of degree $\leq n$, plays a key role [36]. Their $L_p$ analogues have been investigated in a variety of contexts. Here we establish suitable variations for our purposes. We shall find it useful, especially in estimating largest zeros of orthogonal polynomials, to have analogues for exponentials of potentials, a special case of which are absolute values of polynomials. Accordingly, for $T > 0$, we define $\mathbb{P}_T$ to be the set of all functions of the form

$$P(z) = c \exp\left( \int \log |z - x| \, d\nu(x) \right), \tag{4.1}$$

where $c \in \mathbb{R}$, $\nu$ is a measure of total mass $\leq T$ with compact support. In particular, if $P$ is a polynomial of degree $\leq N$, then $|P| \in \mathbb{P}_N$.

**Theorem 4.1.** *Let $\{Q_n\} \in \mathcal{Q}$. Let $T > 0$, $n \geq 1$, $0 < p \leq \infty$ and $P \in \mathbb{P}_{nT-\frac{2}{p}} \setminus \{0\}$. Then*

$$\left\| Pe^{-nQ_n} \right\|_{L_p(I_n \setminus [a_{-n,T}, a_{n,T}])} < \left\| Pe^{-nQ_n} \right\|_{L_p[a_{-n,T}, a_{n,T}]}, \tag{4.2}$$

*and if $p < \infty$,*

$$\left\| Pe^{-nQ_n} \right\|_{L_p(I_n)} < 2^{1/p} \left\| Pe^{-nQ_n} \right\|_{L_p[a_{-n,T}, a_{n,T}]}. \tag{4.3}$$

*In particular, these hold for not-identically vanishing polynomials $P$ of degree $\leq nT - \frac{2}{p}$. For $p = \infty$, (4.3) holds with $<$ replaced by $\leq$.*

© The Author(s) 2018
E. Levin, D.S. Lubinsky, *Bounds and Asymptotics for Orthogonal Polynomials for Varying Weights*, SpringerBriefs in Mathematics,
https://doi.org/10.1007/978-3-319-72947-3_4

For $T = 1$, $0 < p < \infty$, and polynomials $P$ of degree $\leq n - \frac{2}{p}$, $P \neq 0$, Theorem 4.1 gives

$$\left\| Pe^{-nQ_n} \right\|_{L_p(I_n \setminus [-1,1])} < \left\| Pe^{-nQ_n} \right\|_{L_p[-1,1]} .$$

We also prove:

**Theorem 4.2.** *Let* $\{Q_n\} \in \mathcal{Q}$. *Let* $T \in \left[\frac{1}{2}, 2\right]$, $S \geq 0$ *and* $0 < p \leq \infty$. *Then there exists* $C_1, C_2, C_3, \eta > 0$ *such that for* $n \geq 1$ *and* $P \in \mathbb{P}_{nT+S}$,
*(a)*

$$\left\| Pe^{-nQ_n} \right\|_{L_p(I_n \setminus [a_{-n,T}, a_{n,T}])} \leq e^{C_1 (S+1) n^{-1/2}} \left\| Pe^{-nQ_n} \right\|_{L_p[a_{-n,T}, a_{n,T}]} . \tag{4.4}$$

*(b) If* $R_1 \geq C_2 (S + 1)$,

$$\left\| Pe^{-nQ_n} \right\|_{L_p\left(I_n \setminus \left[a_{-n,T} - \frac{R_1}{n}, a_{n,T} + \frac{R_1}{n}\right]\right)} < \left\| Pe^{-nQ_n} \right\|_{L_p[a_{-n,T}, a_{n,T}]} . \tag{4.5}$$

*(c) Uniformly for* $C_2 n^{-1/3} (S + 1) \leq R \leq n^{2/3} \eta$,

$$\left\| Pe^{-nQ_n} \right\|_{L_p\left(I_n \setminus \left[a_{-n,T} - \frac{R}{n^{2/3}}, a_{n,T} + \frac{R}{n^{2/3}}\right]\right)}$$
$$\leq \exp\left(-C_3 R^{3/2}\right) \left\| Pe^{-nQ_n} \right\|_{L_p[a_{-n,T}, a_{n,T}]} .$$

*The constants* $C_1, C_2, C_3, \eta$ *are independent of* $R, R_1, P, n, S$, *but depend on* $T$ *and* $p$.

**Theorem 4.3.** *Let* $\{Q_n\} \in \mathcal{Q}$. *Let* $R > 0$, $S \geq 0$ *and* $0 < p \leq \infty$. *There exist* $C_4, n_0 > 0$, *such that for* $n \geq 1$ *and* $P \in \mathbb{P}_{n+S}$,

$$\left\| Pe^{-nQ_n} \right\|_{L_p(I_n \setminus [-1 + Rn^{-2/3}, 1 - Rn^{-2/3}])}$$
$$\leq C_4 \left\| Pe^{-nQ_n} \right\|_{L_p[-1 + Rn^{-2/3}, 1 - Rn^{-2/3}]} . \tag{4.6}$$

$C_4$ *is independent of* $P, n$, *but depends on* $R, S, p$.

Recall that given an interval $\Delta$, and $x \notin \Delta$, $g_\Delta(z, x)$ denotes the Green's function for $\overline{\mathbb{C}} \setminus \Delta$ with pole at $x$, so that $g_\Delta(z, x) + \log |z - x|$ is harmonic as a function of $z$ in $\overline{\mathbb{C}} \setminus \Delta$ and vanishes on $\Delta$. Moreover, for $x = \infty$, $g_\Delta(z)$ denotes the Green's function with pole at $\infty$. We begin with

**Lemma 4.4.** *Let* $\{Q_n\} \in \mathcal{Q}$. *Let* $T, \Omega > 0$ *and* $0 < p \leq \infty$ *and* $P \in \mathbb{P}_\Omega \setminus \{0\}$. *Represent* $P$ *in the form* (4.1) *with* $c \neq 0$, *so that* $\nu$ *has total mass* $\leq \Omega$. *Let*

$$\Delta = [a_{-n,T}, a_{n,T}] .$$

*(a) For $z \in \mathbb{C} \backslash \Delta$,*

$$|P(z)|^p / \int_\Delta |Pe^{-nQ_n}|^p$$

$$\leq \frac{1}{\pi \, dist\,(z, [-1,1])} \exp\left(-p\left[\int g_\Delta(z, u) dv(u) + n\left(V^{\sigma Q_n, T}(z) - F_{Q_n, T}\right)\right]\right.$$

$$\left. - [p\,(nT - \Omega) - 1]\, g_\Delta(z)\right)$$

$$\leq \frac{1}{\pi \, dist\,(z, [-1,1])} \exp(-p\left[n\left(V^{\sigma Q_n, T}(z) - F_{Q_n, T}\right)\right]$$

$$- [p\,(nT - \Omega) - 1]\, g_\Delta(z)). \tag{4.7}$$

*(b) Moreover, for $z \in \mathbb{C} \backslash \Delta$,*

$$|P(z)| \leq \exp(-n\left(V^{\sigma Q_n, T}(z) - F_{Q_n, T}\right) - (nT - \Omega)\, g_\Delta(z))\, \|\, Pe^{-nQ_n}\,\|_{L_\infty(\Delta)}\,. \tag{4.8}$$

*(c) For $0 < p \leq \infty$,*

$$\left\| P(x)e^{-nQ_n(x)} \exp(-nU_{n,T}\,(x) + \left[nT - \Omega - \frac{2}{p}\right] g_\Delta(x)) \right\|_{L_p(I_n \backslash \Delta)} \tag{4.9}$$

$$\leq \|\, Pe^{-nQ_n}\,\|_{L_p(\Delta)}\,.$$

*Proof.* This is very similar to that of Lemma 9.6 in [25, p. 260], but we provide the details.

(a) We may assume that in $P$, we have $c = 1$ and $v$ has total mass exactly $\Omega$. Let

$$G(z) := \frac{1}{n} \int \{\log |z - u| + g_\Delta(z, u)\}\, dv(u)$$

$$+ \left(V^{\sigma Q_n, T}(z) - F_{Q_n, T}\right) + \left(T - \frac{\Omega}{n}\right) g_\Delta(z).$$

This is harmonic in $\mathbb{C} \backslash \Delta$, and as $z \to \infty$,

$$G(z) = \frac{\Omega}{n} \log |z| + \frac{\Omega}{n} \int g_\Delta\,(\infty, u)\, dv(u) - T \log |z|$$

$$- F_{Q_n, T} + \left(T - \frac{\Omega}{n}\right) \log |z| + \upsilon\,(1)$$

$$= \frac{\Omega}{n} \int g_\Delta\,(\infty, u)\, dv\,(u) - F_{Q_n, T} + o\,(1)\,.$$

Thus it has a finite limit at $\infty$, so is harmonic there too, and thus is harmonic in simply connected $\overline{\mathbb{C}}\backslash\Delta$. Hence it has a single valued harmonic conjugate there, $\widetilde{G}(z)$ say. Then the function

$$f(z) := \exp(G(z) + i\widetilde{G}(z))$$

is analytic and nonvanishing in $\overline{\mathbb{C}}\backslash\Delta$. We can then choose a single valued branch of $f^{np}$ in $\overline{\mathbb{C}}\backslash\Delta$. Letting $\widetilde{g}_\Delta(z)$ denote a harmonic conjugate of $g_\Delta(z)$ in $\mathbb{C}\backslash\Delta$, we have that

$$A(z) := \exp(g_\Delta(z) + i\widetilde{g}_\Delta(z))$$

is analytic in $\mathbb{C}\backslash\Delta$ with a simple pole at $\infty$. More precisely,

$$|A(z)| = \frac{4}{a_{n,T} - a_{-n,T}}|z|(1 + o(1)), \quad z \to \infty.$$

Hence we may apply Cauchy's integral formula for the exterior of a segment to deduce that for $z \notin \Delta$,

$$f^{np}(z)/A(z) = \frac{1}{2\pi i} \int_\Delta \frac{(f^{np}/A)(x + i0) - (f^{np}/A)(x - i0)}{x - z} dx \qquad (4.10)$$

where the terms in the integrand denote boundary values from the upper and lower half planes. Formally, this is derived by taking a contour enclosing $\Delta$, traversed clockwise, and shrinking it to the segment. Next, for a.e. $x \in \Delta$,

$$|(f^{np}/A)(x \pm i0)| = \exp(npG(x)) = |P(x)|^p e^{-npQ_n(x)},$$

recall that $g_\Delta(z, x) = g_\Delta(x) = 0$ for $x \in \Delta^o$. Also, for $z \notin \Delta$,

$$|f^{np}/A|(z) = \exp(npG(z) - g_\Delta(z))$$

$$= |P(z)|^p \exp\left(p\left[\int g_\Delta(z, u)dv(u) + n\left(V^{\sigma Q_n, T}(z) - F_{Q_n, T}\right)\right]\right.$$

$$\left. + [p(nT - \Omega) - 1]g_\Delta(z)\right).$$

Estimating (4.10) in the obvious way and using these last identities together with $g_\Delta(z, u) \geq 0$ yields (4.7).

(b) This follows by taking $p$th roots in (4.7) and letting $p \to \infty$.

(c) We apply the case $p = 2$ of Lemma 4.3 in [25, p. 98] to the function $f^{np/2}/A$ (with $p$ in a different sense to the quoted Lemma 4.3, and not necessarily 2)

$$\left\| f^{np/2}/A \right\|_{L_2(\mathbb{R} \backslash \Delta)} \leq \frac{1}{2} \left( \left\| \left( f^{np/2}/A \right)_+ \right\|_{L_2(\Delta)} + \left\| \left( f^{np/2}/A \right)_- \right\|_{L_2(\Delta)} \right)$$

$$= \left\| Pe^{-nQ_n} \right\|_{L_p(\Delta)}^{p/2}.$$

Hence

$$\left\| P(x) \exp \left( \left[ \int g_\Delta(x,u) d\nu(u) + n \left( V^{\sigma Q_n,T}(x) - F_{Q_n,T} \right) \right] \right. \right. \tag{4.11}$$

$$\left. \left. + \left[ nT - \Omega - \frac{2}{p} \right] g_\Delta(x) \right) \right\|_{L_p(\mathbb{R} \backslash \Delta)}$$

$$\leq \left\| Pe^{-nQ_n} \right\|_{L_p(\Delta)}.$$

Since $g_\Delta(x,u) \geq 0$, and $V^{\sigma Q_n,T}(x) - F_{Q_n,T} = -U_{n,T}(x) - Q_n$ in $I_n$, (4.9) follows.   □

*Proof of Theorem* 4.1. Firstly (4.2) follows from (4.9) and the fact that $\Omega \leq nT - \frac{2}{p}$ and $g_\Delta > 0$ outside $\Delta$, while $U_{n,T} \leq 0$. Next, (4.3) follows from (4.2).   □

*Proof of Theorem* 4.2(a). We apply (4.9) with $\Omega = nT + S$ and $\Delta = [a_{-n,T}, a_{n,T}]$,

$$\left\| P(x) e^{-nQ_n(x)} \exp \left( -\left\{ nU_{n,T}(x) + \left[ S + \frac{2}{p} \right] g_\Delta(x) \right\} \right) \right\|_{L_p(I_n \backslash \Delta)} \tag{4.12}$$

$$\leq \left\| Pe^{-nQ_n} \right\|_{L_p(\Delta)}.$$

Here Lemma 3.6(c) with $B = S + \frac{2}{p}$ gives

$$\sup_{x \in I_n} \left( nU_{n,T}(x) + \left[ S + \frac{2}{p} \right] g_\Delta(x) \right) \leq C \left( S + \frac{2}{p} \right) n^{-1/2} < \infty.$$

□

*Proof of Theorem* 4.2(b),(c).  (b) We use (4.12), which is applicable as $P \in \mathbb{P}_{nT+S}$. We need to show that there exists fixed $R_1$ such that for $x \geq a_{n,T} + \frac{R_1}{n}$, with $x \in I_n$,

$$nU_{n,T}(x) + \left[ S + \frac{2}{p} \right] g_\Delta(x) < 0.$$

Let $\eta_2$ be as in Lemma 3.6. Let $\rho = a_{n,T} + \frac{R_1}{n} \in [a_{n,T}, a_{n,T} + \eta_2]$. For $x \in [\rho, d_n)$, and some $C_1, C_2$ independent of $n, T, S$, (3.35) of Lemma 3.6(b) shows that

$$
\begin{aligned}
nU_{n,T}(x) + &\left[S + \frac{2}{p}\right] g_{[a_{-n,T}, a_{n,T}]}(x) \\
&\leq -C_1 n (\rho - a_{n,T})^{3/2} + \left[S + \frac{2}{p}\right] C_2 (\rho - a_{n,T})^{1/2} \\
&= (\rho - a_{n,T})^{1/2} \left\{ -C_1 n (\rho - a_{n,T}) + \left[S + \frac{2}{p}\right] C_2 \right\} \\
&\leq (\rho - a_{n,T})^{1/2} \left\{ -C_1 R_1 + \left[S + \frac{2}{p}\right] C_2 \right\} < 0,
\end{aligned}
$$

if $R_1 > \left[S + \frac{2}{p}\right] C_2 / C_1$.

(c) Let $\eta_2$ be as in Lemma 3.6. Next, if $\rho = a_{n,T} + \frac{R}{n^{2/3}} \leq a_{n,T} + \eta$, and $x \in [\rho, d_n)$, (3.35) shows that

$$
\begin{aligned}
nU_{n,T}(x) + &\left[S + \frac{2}{p}\right] g_{[a_{-n,T}, a_{n,T}]}(x) \\
&\leq -C_1 R^{3/2} + \left[S + \frac{2}{p}\right] C_2 \left(\frac{R}{n^{2/3}}\right)^{1/2} \\
&= R^{1/2} \left( -C_1 R + \left[S + \frac{2}{p}\right] C_2 n^{-1/3} \right) \\
&\leq -\frac{C_1}{2} R^{3/2},
\end{aligned}
$$

for $R \geq 2 \left[S + \frac{2}{p}\right] \frac{C_2}{C_1} n^{-1/3}$, so from (4.12),

$$
\begin{aligned}
\left\| P(x) e^{-nQ_n(x)} \right\|_{L_p\left(I \setminus \left[a_{-n,T} - \frac{R}{n^{2/3}}, a_{n,T} + \frac{R}{n^{2/3}}\right]\right)} \\
\leq \exp\left(-\frac{C_1}{2} R^{3/2}\right) \left\| P e^{-nQ_n} \right\|_{L_p[a_{-n,T}, a_{n,T}]}.
\end{aligned}
$$

$\qquad\qquad\qquad\qquad\qquad\qquad\qquad\qquad\qquad\qquad\qquad\qquad\qquad\qquad\qquad\quad \square$

*Proof of Theorem* 4.3. Recall that from (3.27) of Lemma 3.3 that $|a_{-n,T} + 1| + |a_{n,T} - 1| \sim |T - 1|$ for $T$ close to 1. Then given $R > 0$, we can choose $L = L(R)$ and $T = T(R) = 1 - Ln^{-2/3}$ such that

$$
\left[-1 + Rn^{-2/3}, 1 - Rn^{-2/3}\right] \supset [a_{-n,T}, a_{n,T}].
$$

Also, given $S > 0$, write

$$n + S = nT + S_1,$$

where

$$S_1 = S + n(1 - T) \sim n^{1/3}.$$

By Theorem 4.2(a), with $S$ replaced by $S_1$, we have for $P \in \mathbb{P}_{n+S} = \mathbb{P}_{nT+S_1}$,

$$\left\| Pe^{-nQ_n} \right\|_{L_p\left(I_n \setminus [-1+Rn^{-2/3}, 1-Rn^{-2/3}]\right)}$$

$$\leq \left\| Pe^{-nQ_n} \right\|_{L_p\left(I_n \setminus [a_{-n,T}, a_{n,T}]\right)}$$

$$\leq e^{C_1(S_1+1)n^{-1/2}} \left\| Pe^{-nQ_n} \right\|_{L_p[a_{-n,T}, a_{n,T}]}$$

$$\leq C \left\| Pe^{-nQ_n} \right\|_{L_p[-1+Rn^{-2/3}, 1-Rn^{-2/3}]}.$$

$\square$

# Chapter 5
# Bounds for Christoffel Functions

Define the $L_p$ Christoffel function for a given external field $Q_n$ by

$$\lambda_{n,p}\left(e^{-pnQ_n}, x\right) = \inf_{\deg(P) \leq n-1} \frac{\int_{I_n} |P(t)|^p\, e^{-pnQ_n(t)}\, dt}{|P(x)|^p}.$$

Of course $\lambda_{n,2}$ is our usual Christoffel function $\lambda_n$.

We shall prove:

**Theorem 5.1.** *Let $A, B > 0$ and $N = N(n)$ satisfy*

$$|N - n| \leq Bn^{1/3}. \tag{5.1}$$

*(a) For $0 < p < \infty$, $x \in I_n$,*

$$\lambda_{N,p}\left(e^{-pnQ_n}, x\right) e^{pnQ_n(x)} \geq \frac{C_1}{n\sqrt{\max\left\{1 - |x|, n^{-2/3}\right\}}}. \tag{5.2}$$

*(b) For $|x| \leq 1 + An^{-2/3}$, with $x \in I_n$,*

$$\lambda_{N,p}(e^{-pnQ_n}, x)e^{pnQ_n(x)} \leq \frac{C_2}{n\sqrt{\max\left\{1 - |x|, n^{-2/3}\right\}}}. \tag{5.3}$$

*The constants $C_1, C_2$ depend on $A, B, p$ but not on $n, N, x$.*

© The Author(s) 2018
E. Levin, D.S. Lubinsky, *Bounds and Asymptotics for Orthogonal Polynomials for Varying Weights*, SpringerBriefs in Mathematics,
https://doi.org/10.1007/978-3-319-72947-3_5

*Proof of Theorem* 5.1(a).  Let $P$ be a nonzero polynomial of degree $m \leq N - 1$ and represent $|P|$ as in (4.1). From the first inequality in (4.7), with $T = 1$, $\Omega = N - 1$, $z = x + iy$, $x \in I_n$, $\Delta = [-1, 1]$,

$$\frac{\int_{I_n} |P(t)|^p \, e^{-pnQ_n(t)} dt}{|P(x)|^p \, e^{-pnQ_n(x)}} = \frac{\int_{I_n} |P(t)|^p \, e^{-pnQ_n(t)} dt}{|P(z)|^p} \left( \frac{|P(z)|}{|P(x)| \, e^{-nQ_n(x)}} \right)^p$$

$$\geq \pi |y| \exp \left( p \left[ \int g_{[-1,1]}(z, u) dv(u) + n \left( V^{\sigma Q_n}(z) - F_{Q_n} \right) \right. \right.$$

$$\left. \left. + (p(n - N + 1) - 1) g_{[-1,1]}(z) \right] \right) \left( \frac{|P(z)|}{|P(x)| \, e^{-nQ_n(x)}} \right)^p.$$

We claim that

$$\exp \left( p \int g_{[-1,1]}(z, t) dv(t) \right) \left| \frac{P(z)}{P(x)} \right|^p$$

$$= \exp \left( p \int \left\{ g_{[-1,1]}(z, t) + \log | \frac{z - t}{x - t} | \right\} dv(t) \right) \geq 1. \tag{5.4}$$

Indeed for each $t \in [-1, 1]$, (with $z = x + iy$ fixed as above)

$$g_{[-1,1]}(z, t) + \log | \frac{z - t}{x - t} | = 0 + \log |1 + \frac{iy}{x - t}| \geq 0$$

and the left-hand side is superharmonic as a function of $t$ in $\overline{\mathbb{C}} \backslash [-1, 1]$. (Indeed, $g_{[-1,1]}(z, t) + \log |z - t|$ is harmonic as a function of $t$ in $\mathbb{C} \backslash [-1, 1]$, while $\log \frac{1}{|x-t|}$ is superharmonic there. Moreover, the left-hand side has a finite limit at $\infty$.) The minimum principle for superharmonic functions shows that the left-hand side $\geq 0$ for all $t \in \mathbb{C}$. Then (5.4) follows. We deduce that

$$\lambda_{N,p} \left( e^{-pnQ_n}, x \right) e^{pnQ_n(x)} = \inf_{\deg(P) \leq n-1} \frac{\int_{I_n} |P(t)|^p \, e^{-pnQ_n(t)} dt}{|P(x)|^p \, e^{-pnQ_n(x)}}$$

$$\geq \pi |y| \exp(pn[V^{\sigma Q_n}(z) - F_{Q_n} + Q_n(x)]$$

$$+ (p(n - N + 1) - 1) g_{[-1,1]}(z)). \tag{5.5}$$

Letting $U_n = U_{n,1}$ be defined by (3.11), we can continue this as

$$= \pi |y| \exp(pn[V^{\sigma Q_n}(z) - V^{\sigma Q_n}(x)] - pnU_n(x)$$

$$+ (p(n - N + 1) - 1) g_{[-1,1]}(z))$$

$$\geq \pi |y| \exp \left( - Cpn |y| \left[ \sqrt{\max\{1 - |x|, 0\}} + \sqrt{|y|} \right] - pnU_n(x) \right.$$

$$\left. - Cn^{1/3} g_{[-1,1]}(z) \right), \tag{5.6}$$

by Lemma 3.5 and (5.1). Choose

$$y = \frac{1}{n\sqrt{\max\left\{1 - |x|, n^{-2/3}\right\}}}.$$

It is easy to see that

$$n|y|\left[\sqrt{\max\{1 - |x|, 0\}} + \sqrt{|y|}\right] \le C.$$

Moreover, by Lemma 3.6(c),

$$pnU_n(x) + Cn^{1/3}g_{[-1,1]}(x + iy)$$

$$= pnU_n(x) + Cn^{1/3}g_{[-1,1]}(x) + Cn^{1/3}\left(g_{[-1,1]}(x + iy) - g_{[-1,1]}(x)\right)$$

$$\le Cn^{-1/6} + Cn^{1/3}\int_{-1}^{1}\log\left(1 + \left(\frac{y}{x - t}\right)^2\right)\frac{dt}{\pi\sqrt{1 - t^2}}.$$

Suppose that we have proved that this last right-hand side is bounded above. Then we can continue (5.6) as

$$\lambda_{N,p}\left(e^{-pnQ_n}, x\right)e^{pnQ_n(x)} \ge C|y|.$$

Our choice of $y$ gives the result (5.2). It remains to estimate the integral. Let us assume $x \ge 0$. We estimate the integral by

$$n^{1/3}\int_{-1}^{1}\log\left(1 + \left(\frac{y}{x - t}\right)^2\right)\frac{dt}{\pi\sqrt{1 - t}}$$

$$\le Cn^{1/3}\int_{\{t: |x-t| \le |1-t|\}}\log\left(1 + \left(\frac{y}{x - t}\right)^2\right)\frac{dt}{\pi\sqrt{|x - t|}}$$

$$+ Cn^{1/3}\int_{\{t: |x-t| \ge |1-t|\}}\log\left(1 + \left(\frac{y}{1 - t}\right)^2\right)\frac{dt}{\pi\sqrt{|1 - t|}}$$

$$\le Cn^{1/3}|y|^{1/2}\int_{-\infty}^{\infty}\log\left(1 + \frac{1}{s^2}\right)\frac{ds}{\sqrt{|s|}} \le C,$$

by the substitutions $x - t = s|y|$ and $1 - t = s|y|$ in the first and second integrals.  □

For $n \ge m \ge 1$ and $x \in I_n$, let

$$\lambda_{m,\infty}(e^{-nQ_n}, x) = \inf_{\deg(P) \le m}\frac{\left\|Pe^{-nQ_n}\right\|_{L_\infty(I_n)}}{|P(x)|}.$$

(The degree is different from the $m-1$ in $\lambda_{m,p}$.) Our upper bounds are based on:

**Lemma 5.2.** *Let* $0 < p < \infty$, $A > 0$, *and* $\{Q_n\} \in \mathcal{Q}$.
(a) *For* $n \geq N > m \geq 1$ *and* $x \in I_n$,

$$\lambda_{N,p}(e^{-pnQ_n}, x)e^{pnQ_n(x)}$$

$$\leq C\left[\lambda_{m,\infty}(e^{-nQ_n}, x)e^{nQ_n(x)}\right]^p \frac{1}{N-m} \max\left\{\sqrt{|1-|x||}, \frac{1}{N-m}\right\}. \tag{5.7}$$

*Here* $C \neq C(n, N, m, x)$.
(b) *Suppose that* $|x| \leq 1 + An^{-2/3}$ *and that we can choose* $m = m(n, x)$ *and* $N = N(n, x)$ *such that*

$$n \geq N \geq \frac{n}{8} \text{ and } 1 - \frac{m}{N} \geq C \max\left\{n^{-2/3}, 1 - |x|\right\}, \tag{5.8}$$

*and*

$$\lambda_{m,\infty}(e^{-nQ_n}, x)e^{nQ_n(x)} \leq C. \tag{5.9}$$

*Then*

$$\lambda_{N,p}(e^{-pnQ_n}, x)e^{pnQ_n(x)} \leq \frac{C}{n} \max\left\{n^{-2/3}, 1 - |x|\right\}^{-1/2}. \tag{5.10}$$

*Proof.* (a) By the restricted range inequality (4.4) with $T = 1$, and as $N \leq n$,

$$\lambda_{N,p}(e^{-pnQ_n}, x)e^{pnQ_n}(x) = \inf_{P \in \mathcal{P}_{N-1}} \int_{I_n} |Pe^{-nQ_n}|^p (s) \, ds / |Pe^{-nQ_n}|^p(x)$$

$$\leq C \inf_{P \in \mathcal{P}_{N-1}} \int_{-1}^{1} |Pe^{-nQ_n}|^p (s) \, ds / |Pe^{-nQ_n}|^p(x)$$

$$\leq C\left[\lambda_{m,\infty}(e^{-nQ_n}, x) / e^{-nQ_n}(x)\right]^p \inf_{P \in \mathcal{P}_{N-m-1}} \int_{-1}^{1} |P|^p (s) \, ds / |P|^p(x)$$

by definition of $\lambda_{m,\infty}$. By a result of P. Nevai on generalized Christoffel functions, [39, Lemma 5, p. 108; Thm. 13, p. 113],

$$\inf_{P \in \mathcal{P}_l} \int_{-1}^{1} |P|^p(s)ds / |P|^p(x) \leq \frac{C}{l} \max\left\{|1 - |x||^{1/2}, \frac{1}{l}\right\}$$

at least for $x \in [-1, 1]$. Here $C \neq C(l, x)$. Almost the same proof shows that for $x > 1$,

$$\inf_{P \in \mathcal{P}_l} \int_{-1}^{1} |P|^p(s)ds / |P|^p(x) \leq \frac{C}{l^2},$$

so that the above inequality persists in $\mathbb{R}$. Setting $l = N - m$ gives the result.

(b) From (a) and (5.8), (5.9),

$$\lambda_{N,p}(e^{-pnQ_n}, x)e^{pnQ_n(x)}$$

$$\leq C\frac{1}{N-m}\max\left\{\sqrt{\max\{n^{-2/3}, 1-|x|\}}, \frac{1}{N-m}\right\}$$

$$\leq C\frac{1}{N}\frac{1}{\max\{n^{-2/3}, 1-|x|\}}$$

$$\times \max\left\{\sqrt{\max\{n^{-2/3}, 1-|x|\}}, \frac{1}{N\max\{n^{-2/3}, 1-|x|\}}\right\}$$

$$\leq C\frac{1}{n}\frac{1}{\sqrt{\max\{n^{-2/3}, 1-|x|\}}}.$$

$\square$

We need to show that for $m = m(n, x)$ satisfying (5.8),

$$\lambda_{m,\infty}(e^{-nQ_n}, x)e^{nQ_n(x)} \leq C.$$

This can be done using Theorem 9.1 in [24, p. 498]:

**Lemma 5.3.** *Let $g(t)$ be a nonnegative continuous function on $[-1, 1]$ satisfying $\int_{-1}^{1} g = 1$. Suppose also that*

$$g(t) = h(t)\sqrt{1-t^2}, \tag{5.11}$$

*where $h$ is a positive continuous function on $[-1, 1]$ whose modulus of continuity $\omega(h; \cdot)$ satisfies for some $\Gamma > 0$,*

$$\omega(h; t) \leq \Gamma\left(\log\frac{1}{t}\right)^{-1}, \quad t \in (0, 1). \tag{5.12}$$

*Let*

$$G(x) = \int_{-1}^{1} \log|x-t|\, g(t)\, dt.$$

*Given $x_0 \in \mathbb{R}$, there exists a polynomial $P_{m,x_0}$ of degree $\leq m$, such that*

$$|P_{m,x_0}(x)| \leq C_1 e^{mG(x)}, \quad x \in \mathbb{R},$$

*and*

$$|P_{m,x_0}(x_0)| \geq C_2 e^{mG(x_0)}.$$

*The constants $C_1$ and $C_2$ depend on $\Gamma$ and the maximum and minimum of h in $[-1, 1]$, but not on $m, x, x_0$, or the particular g or h.*

Now we prove:

**Lemma 5.4.** *Suppose that $A > 0$ and that $x \in I_n$ with $|x| \le 1 + An^{-2/3}$ with $x \in I_n$. Let $B > 0$ and*

$$n \ge N \ge n - Bn^{1/3}. \tag{5.13}$$

*Then we can choose for $n \ge n_0$, $m = m(n, x)$ such that*

$$1 - \frac{m}{N} \ge C \max \left\{ n^{-2/3}, 1 - |x| \right\}, \tag{5.14}$$

*and*

$$\lambda_{m,\infty}(e^{-nQ_n}, x)e^{nQ_n(x)} \le C. \tag{5.15}$$

*Here $n_0$ and C are independent of x.*

*Proof.* We consider several different ranges of $x$. Let $r_0$ be as in Theorem 3.1.

(I) Assume first that $x \in [a_{n,r_0}, 1 - Ln^{-2/3}]$.

Here $L$ is some large number. Choose $r \in [r_0, 1)$ with $x = a_{n,r}$ (this is uniquely possible as $a_{n,r}$ is a strictly increasing continuous function of $r$). Let $m = [nr]$, so that

$$nr - 1 < m \le nr. \tag{5.16}$$

By (3.26) of Lemma 3.3,

$$Ln^{-2/3} \le 1 - x = 1 - a_{n,r} \le C_1(1 - r) \le C_1 \left(1 - \frac{m}{n}\right). \tag{5.17}$$

Here $C_1$ is independent of $m, n, r, L, B$. We assume $L$ is so large that $L/C_1 \ge 2B$, with $B$ as in (5.13). Then

$$n - N \le Bn^{1/3} \le \frac{1}{2}(L/C_1) n^{1/3} \le \frac{1}{2}(n - m)$$

by (5.17). Then also

$$1 - \frac{m}{N} = \frac{1}{N}((n - m) - (n - N)) \ge \frac{n - m}{2N}$$

$$\ge C_1 \left(1 - \frac{m}{n}\right) \ge C_1(1 - x),$$

by (5.17). Then (5.14) follows in this case. Now we can apply Lemma 5.3 to $g = \hat{\sigma}_{Q_n,r}$: note that the smoothness condition (5.12) is satisfied (uniformly in $n$ and $r$), by (3.14) of Theorem 3.1. So there exists a polynomial $P_{m,x}$ of degree $\leq m$, such that

$$|P_{m,x}(t)| \leq C_1 \exp\left(-mV^{\hat{\sigma}_{Q_n,r}}(t)\right), \quad t \in \mathbb{R}$$

and

$$|P_{m,x}(x)| \geq C_2 \exp\left(-mV^{\hat{\sigma}_{Q_n,r}}(x)\right).$$

Next from the first equation in (3.9),

$$
\begin{aligned}
-V^{\hat{\sigma}_{Q_n,r}}(L_{n,r}(y)) &= \int_{-1}^{1} \log |L_{n,r}(y) - u|\, \hat{\sigma}_{Q_n,r}(u)\, du \\
&= \frac{\delta_{n,r}}{r} \int_{-1}^{1} \log |L_{n,r}(y) - u|\, \sigma_{Q_n,r}\left(L_{n,r}^{[-1]}(u)\right) du \\
&= \frac{1}{r} \int_{a-n,r}^{a_{n,r}} \log |L_{n,r}(y) - L_{n,r}(s)|\, \sigma_{Q_n,r}(s)\, ds \\
&= \frac{1}{r} \int_{a-n,r}^{a_{n,r}} \log |y - s|\, \sigma_{Q_n,r}(s)\, ds - \log \delta_{n,r} \\
&= -\frac{1}{r} V^{\sigma_{Q_n,r}}(y) - \log \delta_{n,r}.
\end{aligned}
$$

Now let

$$R_n(t) = P_{m,L_{n,r}(x)}(L_{n,r}(t)), \quad t \in \mathbb{R}. \tag{5.18}$$

Then for $t \in I_n$, and as $n \geq m/r$, the inequalities above give

$$
\begin{aligned}
\left| R_n(t)\, e^{-nQ_n(t)} \right| &\leq C_1 \exp\left(-mV^{\hat{\sigma}_{Q_n,r}}(L_{n,r}(t)) - \frac{m}{r}Q_n(t)\right) \\
&= C_1 \exp\left(-\frac{m}{r}[V^{\sigma_{Q_n,r}}(t) + r\log \delta_{n,r}] - \frac{m}{r}Q_n(t)\right) \\
&= C_1 \exp\left(\frac{m}{r}U_{n,r}(t) - \frac{m}{r}F_{Q_n,r} - m\log \delta_{n,r}\right) \\
&\leq C_1 \exp\left(-\frac{m}{r}F_{Q_n,r} - m\log \delta_{n,r}\right) = C_1 A_n, \tag{5.19}
\end{aligned}
$$

say. Here we are using $U_{n,r} \leq 0$. Also, for the given $x = a_{n,r}$,

$$
\begin{aligned}
\left| R_n(x) e^{-nQ_n(x)} \right| \\
\geq C_2 \exp\left( -mV^{\hat{\sigma}_{Q_{n,r}}}(L_{n,r}(x)) - \frac{m}{r}Q_n(x) \right) \exp\left( \left( \frac{m}{r} - n \right) Q_n(x) \right) \\
\geq C_2 \exp\left( \frac{m}{r}U_{n,r}(x) - \frac{m}{r}F_{Q_{n,r}} - m\log\delta_{n,r} \right) \exp\left( -\frac{1}{r}Q_n(x) \right) \\
= C_2 A_n \exp\left( -\frac{1}{r}Q_n(x) \right) \geq C_3 A_n.
\end{aligned}
\tag{5.20}
$$

Here we have used that $U_{n,r}(a_{n,r}) = 0$. Then $R_n$ is a polynomial of degree $\leq m$, and

$$
\lambda_{m,\infty}(e^{-nQ_n}, x)e^{nQ_n(x)} \leq \frac{\left\| R_n e^{-nQ_n} \right\|_{L_\infty(I_n)}}{\left| R_n(x) e^{-nQ_n(x)} \right|} \leq \frac{C_3}{C_1}.
\tag{5.21}
$$

(II) Assume next that $x \in [-1 + Ln^{-2/3}, a_{-n,r_0}]$.
    This is similar to (I).
(III) Assume next that $x \in (a_{-n,r_0}, a_{n,r_0})$.
    Recall from Lemma 3.3(b) that $1 - |a_{\pm n,r_0}| \geq C > 0$. Thus also $1 - |x| \geq C > 0$.
Choose $r = \frac{3}{4}$ and $m = \left[ \frac{3}{4}n \right]$. We have

$$
1 - \frac{m}{N} \sim 1 \sim \max\left\{ n^{-2/3}, 1 - |x| \right\},
$$

so still (5.14) holds, and we proceed as in (I), choosing $R_n$ by (5.18).
(IV) Assume next that $x \in [1 - Ln^{-2/3}, 1 + An^{-2/3}]$.
    In this case, we choose

$$
m = n - \left[ (A + B) n^{1/3} \right] \leq N - An^{1/3} + 1,
$$

so that

$$
1 - \frac{m}{N} \geq \frac{An^{1/3} - 1}{N} \geq Cn^{-2/3} \geq C\max\left\{ 1 - |x|, n^{-2/3} \right\},
$$

and (5.14) holds. We choose

$$
r = \frac{m}{n} = 1 - \frac{\left[ (A + B) n^{1/3} \right]}{n}
$$

so that $m = rn$. We choose $R_n$ by (5.18). Exactly as at (5.19), for $t \in I_n$,

$$\left| R_n(t) e^{-nQ_n(t)} \right| \leq C_1 A_n.$$

Moreover, as at (5.20),

$$\left| R_n(x) e^{-nQ_n(x)} \right| \geq C_2 A_n \exp\left( \frac{m}{r} U_{n,r}(x) \right) \exp\left( -\frac{1}{r} Q_n(x) \right).$$

Here $\frac{1}{r} Q_n(x) \leq 2Q_n(x) \leq C$, while if $x \leq a_{n,r}$, we have $U_{n,r}(x) = 0$ and if $x > a_{n,r}$, then

$$x - a_{n,r} \leq 1 + An^{-2/3} - a_{n,r} \leq C(1 - r) + An^{-2/3} \leq Cn^{-2/3}$$

so from (3.34) of Lemma 3.6(b),

$$\frac{m}{r} U_{n,r}(x) \geq -Cn(x - a_{n,r})^{3/2} \geq -C.$$

Thus

$$\left| R_n(x) e^{-nQ_n(x)} \right| \geq C_3.$$

Then (5.21) follows.

(V) Assume next that $x \in [-1 - An^{-2/3}, -1 + Ln^{-2/3}]$.

This is similar to (IV). $\qquad\qquad\square$

*Proof of Theorem* 5.1(b). This follows from Lemmas 5.2 and 5.4 for the case $n \geq N \geq n - Bn^{1/3}$. Since $\lambda_{n,p}$ is decreasing in $n$, we also obtain the bound for the full range (5.1). $\qquad\qquad\square$

# Chapter 6
# Spacing of Zeros

Recall that we order the zeros of $p_{n,n}$ as

$$x_{nn} < x_{n-1,n} < \cdots < x_{2n} < x_{1n}.$$

We prove

**Theorem 6.1.** *Assume* $\{Q_n\} \in \mathcal{Q}$.
(a)

$$1 - \frac{C_2}{n^{2/3}} \leq x_{1n} < a_{n,1+\frac{1}{2n}} \leq 1 + \frac{C_2}{n}. \tag{6.1}$$

*Analogous inequalities hold for* $x_{nn}$.
(b) *For some* $C > 0$, *and* $2 \leq j \leq n$,

$$x_{j-1,n} - x_{jn} \leq \frac{C}{n\sqrt{\max\{1 - |x_{jn}|, n^{-2/3}\}}}. \tag{6.2}$$

We note that further results on spacing of zeros appear in Theorems 13.5 and 14.2.

*Proof of the Right Inequality in* (6.1). We use

$$1 - \frac{x_{1n}}{a_{n,1+\frac{1}{2n}}} = \min_{P \in \mathcal{P}_{n-1}} \frac{\int_{I_n} \left(1 - \frac{x}{a_{n,1+\frac{1}{2n}}}\right) P^2(x) e^{-2nQ_n(x)} dx}{\int_{I_n} P^2(x) e^{-2nQ_n(x)} dx}. \tag{6.3}$$

This is an easy consequence of the Gauss quadrature formula, see for example [49, p. 188]. The minimum is attained when we take $P$ to be the fundamental polynomial

© The Author(s) 2018

E. Levin, D.S. Lubinsky, *Bounds and Asymptotics for Orthogonal Polynomials*
*for Varying Weights*, SpringerBriefs in Mathematics,
https://doi.org/10.1007/978-3-319-72947-3_6

$\ell_{1n} \in \mathcal{P}_{n-1}$ of Lagrange interpolation satisfying $\ell_{1n}(x_{jn}) = \delta_{1j}$. Here if $P$ is a nonzero polynomial of degree $\leq n - 1$, then $\left|1 - \frac{x}{a_{n,1+\frac{1}{2n}}}\right|^{1/2} |P(x)| \in \mathbb{P}_{n-\frac{1}{2}}$, so (4.2) of Theorem 4.1 with $T = 1 + \frac{1}{2n}, p = 2, nT - \frac{2}{p} = n - \frac{1}{2}$, gives that

$$\int_{I_n \setminus \left[a_{-n,1+\frac{1}{2n}}, a_{n,1+\frac{1}{2n}}\right]} \left|1 - \frac{x}{a_{n,1+\frac{1}{2n}}}\right| P(x)^2 \, e^{-2nQ_n(x)} dx$$

$$< \int_{\left[a_{-n,1+\frac{1}{2n}}, a_{n,1+\frac{1}{2n}}\right]} \left|1 - \frac{x}{a_{n,1+\frac{1}{2n}}}\right| P(x)^2 \, e^{-2nQ_n(x)} dx.$$

Then

$$\int_{I_n} \left(1 - \frac{x}{a_{n,1+\frac{1}{2n}}}\right) P^2(x) e^{-2nQ_n(x)} dx > 0,$$

so that from (6.3),

$$1 - \frac{x_{1n}}{a_{n,1+\frac{1}{2n}}} > 0.$$

Finally, $a_{n,1+\frac{1}{2n}} < 1 + \frac{C}{n}$ by (3.26) of Lemma 3.3(a). $\qquad \square$

For the left inequality in (6.1), we need:

**Lemma 6.2.** *Let $A > 0$. There exist polynomials $R_n$ of degree $\leq 2n - 2n^{1/3}$ such that*

$$R_n(x) e^{-2nQ_n(x)} \sim \sqrt{\max\{1 - |x|, n^{-2/3}\}}, \quad \text{for } x \in I_n \text{ with } |x| \leq 1 + An^{-2/3}.$$

*Proof.* Define

$$N = N(n) = n - \left[n^{1/3}\right].$$

Let

$$R_n(x) = \frac{1}{n} \lambda_{N,2}^{-1} \left(e^{-2nQ_n}, x\right),$$

a polynomial of degree $\leq 2N - 2 \leq 2n - 2n^{1/3}$. By Theorem 5.1(b), for $|x| \leq 1 + An^{-2/3}$,

$$R_n(x) e^{-2nQ_n(x)} \geq C \sqrt{\max\{1 - |x|, n^{-2/3}\}}$$

while using monotonicity of Christoffel functions in the degree, and Theorem 5.1(a),

$$R_n(x) e^{-2nQ_n(x)} \le \frac{1}{n} \lambda_{n,2}^{-1} \left( e^{-2nQ_n}, x \right) e^{-2nQ_n(x)} \le C \sqrt{\max\{1 - |x|, n^{-2/3}\}}.$$

$\square$

*Proof of the Left Inequality in (6.1).* We use (6.3) in the slightly modified form

$$1 - x_{1n} = \min_{P \in \mathcal{P}_{2n-2}, P \ge 0 \text{ in } I_n} \frac{\int_{I_n} (1 - x) P(x) e^{-2nQ_n(x)} dx}{\int_{I_n} P(x) e^{-2nQ_n(x)} dx}$$

$$\le \min_{P \in \mathcal{P}_{2n-2}, P \ge 0 \text{ in } I_n} \frac{\int_{I_n} |1 - x| P(x) e^{-2nQ_n(x)} dx}{\int_{I_n} P(x) e^{-2nQ_n(x)} dx}$$

$$\le C \min_{P \in \mathcal{P}_{2n-2}, P \ge 0 \text{ in } I_n} \frac{\int_{-1}^{1} (1 - x) P(x) e^{-2nQ_n(x)} dx}{\int_{-1}^{1} P(x) e^{-2nQ_n(x)} dx},$$

by (4.4) with $T = 1$, $p = 1$, and $Q_n$ replaced by $2Q_n$. Now choose $P = R_n S^4$, where $R_n$ is as in Lemma 6.2 and

$$S(x) = \frac{T_\ell(x) - 1}{x - 1}$$

and $\ell$ is the largest even integer $\le \frac{1}{8} n^{1/3} - 1$. Here $T_\ell$ is the classical Chebyshev polynomial of degree $\ell$. Note that for $x \in [-1, 1]$, Bernstein's Inequality gives

$$|S(x)| \le C \min \left\{ \ell^2, \frac{1}{|x - 1|} \right\}$$

and for some $\eta > 0$,

$$S(x) \ge \frac{1}{2} \ell^2, \quad x \in \left[1 - \eta \ell^{-2}, 1\right].$$

Lemma 6.2 and these last inequalities give

$$1 - x_{1n} \le C \frac{\int_{-1}^{1} (1 - x) \sqrt{\max\{1 - |x|, n^{-2/3}\}} \min\left\{\ell^2, \frac{1}{|x-1|}\right\}^4 dx}{\int_{1-\eta\ell^{-2}}^{1} \sqrt{\max\{1 - |x|, n^{-2/3}\}} \ell^8 dx}$$

$$\le C \ell^{-5} \int_{-1}^{1} (1 - x) \sqrt{\max\{1 - x, \ell^{-2}\}} \min\left\{\ell^2, \frac{1}{|x-1|}\right\}^4 dx$$

$$= C \ell^{-2} \int_{0}^{2\ell^2} s \sqrt{\max\{s, 1\}} \min\left\{1, \frac{1}{s}\right\}^4 ds \le C n^{-2/3},$$

by the substitution $1 - x = \ell^{-2} s$.

$\square$

For the spacing, we use:

**Lemma 6.3.** *Let*

$$a \leq y_1 < y_2 < \cdots < y_m \leq b.$$

*Assume* $\{\ell_j\}_{j=1}^m$ *are the fundamental polynomials of Lagrange interpolation at* $\{y_j\}$. *Let* $w : (a, b) \to (0, \infty)$ *and assume that* $q = \log \frac{1}{w}$ *is such that* $q'$ *exists and is nondecreasing in* $(a, b)$. *Then for* $1 \leq j \leq m - 1$,

$$\ell_j(x) w^{-1}(y_j) w(x) + \ell_{j+1}(x) w^{-1}(y_{j+1}) w(x) \geq 1, \quad x \in [y_j, y_{j+1}].$$

*Proof.* See [25, Lemma 11.8, p. 320].                                        □

*Proof of Theorem 6.1(b).* Assume that $x_{jn}$ and $x_{j-1,n} \geq 0$. Let $\lambda_{jn} = \lambda_n(e^{-2nQ_n}, x_{jn})$. We have

$$\lambda_{jn} e^{2nQ_n(x_{jn})} + \lambda_{j-1,n} e^{2nQ_n(x_{j-1,n})}$$

$$= \int_{I_n} \left( \ell_{jn}^2(x) e^{2nQ_n(x_{jn})} + \ell_{j-1,n}^2(x) e^{2nQ_n(x_{j-1,n})} \right) e^{-2nQ_n(x)} dx$$

$$\geq \frac{1}{2} \int_{x_{jn}}^{x_{j-1,n}} \left( \ell_{jn}(x) e^{nQ_n(x_{jn})} + \ell_{j-1,n}(x) e^{nQ_n(x_{j-1,n})} \right)^2 e^{-2nQ_n(x)} dx$$

$$\geq \frac{1}{2} \left( x_{j-1,n} - x_{jn} \right),$$

by Lemma 6.3. Assume now that $x_{jn}$ and $x_{j-1,n} \geq 0$. We apply our upper bounds (5.3) for Christoffel functions:

$$x_{j-1,n} - x_{jn} \leq \frac{C}{n} \left( \frac{1}{\sqrt{\max\{1 - x_{jn}, n^{-2/3}\}}} + \frac{1}{\sqrt{\max\{1 - x_{j-1,n}, n^{-2/3}\}}} \right)$$

$$\leq \frac{2C}{n} \frac{1}{\sqrt{\max\{1 - x_{j-1,n}, n^{-2/3}\}}}. \tag{6.4}$$

We claim that uniformly in $j, n$,

$$\max\{1 - x_{j-1,n}, n^{-2/3}\} \sim \max\{1 - x_{jn}, n^{-2/3}\}. \tag{6.5}$$

If $x_{jn}$ and $x_{j-1,n} \geq 0$ and

$$\max\{1 - x_{jn}, n^{-2/3}\} \leq 2 \max\{1 - x_{j-1,n}, n^{-2/3}\},$$

this is true. If this last inequality fails, then

$$1 - x_{jn} > 2 \left(1 - x_{j-1,n}\right),$$

so

$$x_{j-1,n} - x_{jn} = \left(1 - x_{jn}\right) - \left(1 - x_{j-1,n}\right) \geq \frac{1}{2} \left(1 - x_{jn}\right)$$

so (6.4) gives

$$1 - x_{jn} \leq 2 \left(x_{j-1,n} - x_{jn}\right) \leq \frac{C}{n} \frac{1}{\sqrt{\max \left\{1 - x_{j-1,n}, n^{-2/3}\right\}}} \leq Cn^{-2/3}.$$

Thus in this case,

$$\max\{1 - x_{jn}, n^{-2/3}\} \sim \max\{1 - x_{j-1,n}, n^{-2/3}\} \sim n^{-2/3}$$

and we again have (6.5) and hence also (6.2). The case where $x_{jn}$ and $x_{j-1,n} \leq 0$ is similar, and the case where $x_{j-1,n} \geq 0 \geq x_{j,n}$ is easily handled as in this case $\max\{1 - x_{jn}, n^{-2/3}\} \sim \max\{1 - x_{j-1,n}, n^{-2/3}\} \sim 1$. $\qquad\square$

# Chapter 7
# Bounds on Orthogonal Polynomials

We prove:

**Theorem 7.1.** *Assume* $\{Q_n\} \in \mathcal{Q}$. *Let* $A > 0$. *For* $n \geq 1$ *and* $m \geq 1$ *with*

$$|m - n| \leq An^{1/3}, \tag{7.1}$$

*(a)*

$$\sup_{x \in I_n} |p_{n,m}(x)| \, e^{-nQ_n(x)} \, \left|1 - x^2\right|^{1/4} \sim 1. \tag{7.2}$$

*(b)*

$$\sup_{x \in I_n} |p_{n,m}(x)| \, e^{-nQ_n(x)} \left[|1 - |x|| + n^{-2/3}\right]^{1/4} \sim 1. \tag{7.3}$$

Most of this chapter will be devoted to proving Theorem 7.1 for $m = n$. We then manipulate the definition of $\{Q_n\}$ to obtain the general case. Our approach follows that in [24, 25].

We begin by defining

$$\bar{Q}_n(x, t) = \frac{Q_n'(x) - Q_n'(t)}{x - t}$$

and

$$A_n(x) = 2n \int_{I_n} p_{n,n}^2(t) \, \bar{Q}_n(x, t) \, e^{-2nQ_n(t)} dt. \tag{7.4}$$

© The Author(s) 2018
E. Levin, D.S. Lubinsky, *Bounds and Asymptotics for Orthogonal Polynomials for Varying Weights*, SpringerBriefs in Mathematics,
https://doi.org/10.1007/978-3-319-72947-3_7

**Lemma 7.2.** *(a)*

$$p'_{n,n}\left(x_{jn}\right) = \frac{\gamma_{n,n-1}}{\gamma_{n,n}} p_{n,n-1}\left(x_{jn}\right) A_n\left(x_{jn}\right).$$  (7.5)

*(b)*

$$\left|p_{n,n}\left(x\right)\right| \leq \left|x - x_{jn}\right| \left[K_n\left(x,x\right) A_n\left(x_{jn}\right)\right]^{1/2}.$$  (7.6)

*Proof.* (a) We use the reproducing kernel relation, and then integrate by parts:

$$
\begin{aligned}
p'_{n,n}\left(x\right) &= \int_{I_n} K_n\left(x,t\right) p'_{n,n}\left(t\right) e^{-2nQ_n(t)} dt \\
&= K_n\left(x,t\right) p_{n,n}\left(t\right) e^{-2nQ_n(t)} \big|_{t=c_n}^{t=d_n} \\
&\quad - \int_{I_n} p_{n,n}\left(t\right) \frac{d}{dt} \left\{ K_n\left(x,t\right) e^{-2nQ_n(t)} \right\} dt \\
&= 0 - \int_{I_n} p_{n,n}\left(t\right) \left( \frac{\partial}{\partial t} K_n\left(x,t\right) \right) e^{-2nQ_n(t)} dt \\
&\quad + 2n \int_{I_n} p_{n,n}\left(t\right) K_n\left(x,t\right) Q'_n\left(t\right) e^{-2nQ_n(t)} dt \\
&= 2n \int_{I_n} p_{n,n}\left(t\right) K_n\left(x,t\right) Q'_n\left(t\right) e^{-2nQ_n(t)} dt.
\end{aligned}
$$

Here we have used the fact that $Q_n\left(x\right) / \log\left(2 + |x|\right)$ has limit $\infty$ at $c_n, d_n$, and also orthogonality. Now we set $x = x_{jn}$ and use the Christoffel-Darboux formula, and then orthogonality again:

$$
\begin{aligned}
p'_{n,n}\left(x_{jn}\right) &= 2n \int_{I_n} p_{n,n}\left(t\right) K_n\left(x_{jn},t\right) Q'_n\left(t\right) e^{-2nQ_n(t)} dt \\
&= 2n \frac{\gamma_{n,n-1}}{\gamma_{n,n}} p_{n,n-1}\left(x_{jn}\right) \int_{I_n} \frac{p^2_{n,n}\left(t\right) Q'_n\left(t\right)}{t - x_{jn}} e^{-2nQ_n(t)} dt \\
&= 2n \frac{\gamma_{n,n-1}}{\gamma_{n,n}} p_{n,n-1}\left(x_{jn}\right) \int_{I_n} p^2_{n,n}\left(t\right) \frac{Q'_n\left(t\right) - Q'_n\left(x_{jn}\right)}{t - x_{jn}} e^{-2nQ_n(t)} dt.
\end{aligned}
$$

(b) First, by the confluent form of the Christoffel-Darboux formula,

$$
\begin{aligned}
K_n\left(x_{jn}, x_{jn}\right) &= \frac{\gamma_{n,n-1}}{\gamma_{n,n}} p'_{n,n}\left(x_{jn}\right) p_{n,n-1}\left(x_{jn}\right) \\
&= \left( \frac{\gamma_{n,n-1}}{\gamma_{n,n}} p_{n,n-1}\left(x_{jn}\right) \right)^2 A_n\left(x_{jn}\right)
\end{aligned}
$$

so that

$$\left| \frac{\gamma_{n,n-1}}{\gamma_{n,n}} p_{n,n-1}\left(x_{jn}\right) \right| = \left(K_n\left(x_{jn},x_{jn}\right)/A_n\left(x_{jn}\right)\right)^{1/2}.$$

Then the Christoffel-Darboux formula gives

$$\left|p_{n,n}\left(x\right)\right| = \left|x - x_{jn}\right| \frac{\left|K_n\left(x,x_{jn}\right)\right|}{\left|\frac{\gamma_{n,n-1}}{\gamma_{n,n}} p_{n,n-1}\left(x_{jn}\right)\right|}$$

$$\leq \left|x - x_{jn}\right| \left\{ \frac{K_n\left(x,x\right)K_n\left(x_{jn},x_{jn}\right)}{K_n\left(x_{jn},x_{jn}\right)/A_n\left(x_{jn}\right)} \right\}^{1/2},$$

by Cauchy-Schwarz, and the identity above. □

Recall that $t_n$ is defined by $Q'_n\left(t_n\right) = 0$, cf. (1.16).

**Lemma 7.3.**
*(a)*

$$\int_{I_n} p_{n,m}^2\left(t\right)e^{-2nQ_n(t)}tQ'_n\left(t\right)dt = \frac{m+\frac{1}{2}}{n}. \tag{7.7}$$

*(b)*

$$\int_{I_n} p_{n,m}^2\left(t\right)e^{-2nQ_n(t)}Q'_n\left(t\right)dt = 0. \tag{7.8}$$

*(c) For $\eta > 0$,*

$$\int_{I_n\setminus(t_n-\eta,t_n+\eta)} p_{n,m}^2\left(t\right)e^{-2nQ_n(t)}\left|Q'_n\left(t\right)\right|dt \leq \frac{m+\frac{1}{2}}{n\eta}. \tag{7.9}$$

*Proof.* (a) An integration by parts gives

$$2n\int_{I_n} p_{n,m}^2\left(t\right)e^{-2nQ_n(t)}tQ'_n\left(t\right)dt$$

$$= -\int_{I_n} p_{n,m}^2\left(t\right)t\frac{d}{dt}\left\{e^{-2nQ_n(t)}\right\}dt$$

$$= -\left[p_{n,m}^2\left(t\right)te^{-2nQ_n(t)}\right]_{t=c_n}^{t=d_n} + \int_{I_n}\frac{d}{dt}\left[p_{n,m}^2\left(t\right)t\right]e^{-2nQ_n(t)}dt$$

$$= \int_{I_n}\left[p_{n,m}^2\left(t\right) + 2p_{n,m}\left(t\right)p'_{n,m}\left(t\right)t\right]e^{-2nQ_n(t)}dt$$

$$= 1 + 2m.$$

(b) As in (a),

$$2n \int_{I_n} p_{n,m}^2 (t) \, e^{-2nQ_n(t)} Q_n' (t) \, dt$$

$$= -\left[ p_{n,m}^2 (t) \, e^{-2nQ_n(t)} \right]_{t=c_n}^{t=d_n} + \int_{I_n} \frac{d}{dt} \left[ p_{n,m}^2 (t) \right] e^{-2nQ_n(t)} dt$$

$$= \int_{I_n} 2 p_{n,m} (t) \, p_{n,m}' (t) \, e^{-2nQ_n(t)} dt = 0.$$

(c) From (a), (b), and as $(t - t_n) \, Q_n' (t) \geq 0, t \in I_n$, so

$$\frac{m + \frac{1}{2}}{n} = \int_{I_n} p_{n,m}^2 (t) \, e^{-2nQ_n(t)} \left| (t - t_n) \, Q_n' (t) \right| dt$$

$$\geq \eta \int_{I_n \setminus (t_n - \eta, t_n + \eta)} p_{n,m}^2 (t) \, e^{-2nQ_n(t)} \left| Q_n' (t) \right| dt.$$

$\square$

*Proof of Theorem 7.1 for $m(n) = n, n \geq 1$.* (a) Let $x \in [-1, 1]$. Choose $j$ such that $x_{jn}$ is the closest zero of $p_{n,n}$ in $[-1, 1]$ to $x$. Then using Theorem 6.1 and (6.5), we see that

$$\left| x - x_{jn} \right| \leq \frac{C}{n \sqrt{\max\{1 - |x|, n^{-2/3}\}}}.$$

Also from (5.2), with $p = 2$ and $N = n$,

$$K_n (x, x) \, e^{-2nQ_n(x)} \leq Cn \sqrt{\max\{1 - |x|, n^{-2/3}\}}.$$

Then from Lemma 7.2(b),

$$\left| p_{n,n} (x) \right| e^{-nQ_n(x)} (1 - x^2)^{1/4}$$

$$\leq C \left( 1 - x^2 \right)^{1/4} \left[ \frac{1}{n \sqrt{\max\{1 - |x|, n^{-2/3}\}}} A_n \left( x_{jn} \right) \right]^{1/2}$$

$$\leq C \left[ \int_{I_n} p_{n,n}^2 (t) \, \bar{Q}_n \left( x_{jn}, t \right) e^{-2nQ_n(t)} dt \right]^{1/2}. \tag{7.10}$$

Let

$$M_n = \sup_{t \in I_n} \left| p_{n,n} (t) \right| e^{-nQ_n(t)} \left| 1 - t^2 \right|^{1/4}. \tag{7.11}$$

In view of Theorem 4.2(a) with $T = 1$, $S = \frac{1}{2}$, and $P(t) = |p_{n,n}(t)| \, |1 - t^2|^{1/4} \in \mathbb{P}_{n+\frac{1}{2}}$, we have

$$M_n \leq C_0 \sup_{t \in [-1,1]} |p_{n,n}(t)| \, e^{-nQ_n(t)} |1 - t^2|^{1/4}. \tag{7.12}$$

Let $\eta > 0$ be a small positive number, independent of $n, x$. We shall choose it to be small enough later. We estimate the integral in (7.10) by splitting $I_n$ into 3 ranges. First,

$$J_1 = \int_{I_n \setminus ([t_n - \eta, t_n + \eta] \cup [x_{jn} - \eta, x_{jn} + \eta])} p_{n,n}^2(t) \, \bar{Q}_n(x_{jn}, t) \, e^{-2nQ_n(t)} dt$$

$$\leq \frac{1}{\eta} \left\{ |Q_n'(x_{jn})| + \int_{I_n \setminus [t_n - \eta, t_n + \eta]} p_{n,n}^2(t) \, |Q_n'(t)| \, e^{-2nQ_n(t)} dt \right\}$$

$$\leq \frac{1}{\eta} \left\{ \|Q_n'\|_{L_\infty[-1,1]} + \frac{2}{\eta} \right\} = C(\eta), \tag{7.13}$$

by (7.9). In view of (3.23), we may assume $\eta$ is so small that for some $C_1 > 0$, we have

$$[t_n - \eta, t_n + \eta] \subset [-1 + C_1, 1 - C_1].$$

Then $\sqrt{1 - t^2}$ is bounded below by a positive constant in $[t_n - \eta, t_n + \eta]$ for all $n \geq 1$, so

$$J_2 = \int_{I_n \cap [t_n - \eta, t_n + \eta]} p_{n,n}^2(t) \, \bar{Q}_n(x_{jn}, t) \, e^{-2nQ_n(t)} dt$$

$$\leq M_n^2 \int_{I_n \cap [t_n - \eta, t_n + \eta]} \bar{Q}_n(x_{jn}, t) \, \frac{dt}{\sqrt{|1 - t^2|}}$$

$$\leq C M_n^2 \int_{t_n - \eta}^{t_n + \eta} |x_{jn} - t|^{\alpha - 1} dt,$$

by (1.11). If $x_{jn} \in [t_n - 2\eta, t_n + 2\eta]$, we estimate in an obvious way to obtain

$$J_2 \leq C_1 M_n^2 \eta_n^\alpha. \tag{7.14}$$

If $x_{jn} \notin [t_n - 2\eta, t_n + 2\eta]$, then as $\alpha < 1$,

$$J_2 \leq C M_n^2 \eta^{\alpha - 1} (2\eta).$$

So in all cases, we have (7.14). Here it is crucial that $C_1$ is independent of $\eta$ (as well as $n, x$). Now we must deal with $I_n \cap [x_{jn} - \eta, x_{jn} + \eta]$. We split this as a union $\mathcal{S}_{1n} \cup \mathcal{S}_{2n}$, where $\mathcal{S}_{1n} \subset I_n \backslash I_0$ and $\mathcal{S}_{2n} \subset I_0$. In $\mathcal{S}_{1n}$, $\sqrt{1 - t^2}$ is bounded below, independently of $n$, so that

$$
\begin{aligned}
J_{31} &= \int_{\mathcal{S}_{1n}} p_{n,n}^2(t) \, \bar{Q}_n(x_{jn}, t) \, e^{-2nQ_n(t)} dt \leq M_n^2 \int_{\mathcal{S}_{1n}} \bar{Q}_n(x_{jn}, t) \, \frac{dt}{\sqrt{|1 - t^2|}} \\
&\leq C M_n^2 \int_{\mathcal{S}_{1n}} \bar{Q}_n(x_{jn}, t) \, dt \\
&\leq C M_n^2 \int_{x_{jn} - \eta}^{x_{jn} + \eta} |x_{jn} - t|^{\alpha - 1} \, dt \leq C_1 M_n^2 \eta^\alpha.
\end{aligned}
\tag{7.15}
$$

Now $\mathcal{S}_{2n}$ consists of points close to either $-1$ or $1$. Assume the latter. Then

$$
\begin{aligned}
J_{32} &= \int_{\mathcal{S}_{2n}} p_{n,n}^2(t) \, \bar{Q}_n(x_{jn}, t) \, e^{-2nQ_n(t)} dt \\
&\leq M_n^2 \int_{\mathcal{S}_{2n}} \bar{Q}_n(x_{jn}, t) \, \frac{dt}{\sqrt{|1 - t^2|}} \\
&\leq C M_n^2 \int_{x_{jn} - \eta}^{x_{jn} + \eta} |x_{jn} - t|^{\alpha_1 - 1} \, \frac{dt}{\sqrt{|1 - t|}} \\
&= C M_n^2 (1 - x_{jn})^{\alpha_1 - 1/2} \int_{1 - \frac{\eta}{1 - x_{jn}}}^{1 + \frac{\eta}{1 - x_{jn}}} |1 - s|^{\alpha_1 - 1} \, \frac{ds}{\sqrt{|s|}}.
\end{aligned}
\tag{7.16}
$$

Here we used the substitution $1 - t = s(1 - x_{jn})$. We now consider two subcases:
(I) $\eta \leq \frac{1}{2} (1 - x_{jn})$

Then the range of integration in (7.16) is contained in $\left[\frac{1}{2}, 2\right]$, so we continue (7.16) as

$$
\begin{aligned}
J_{32} &\leq C M_n^2 (1 - x_{jn})^{\alpha_1 - 1/2} \int_{1 - \frac{\eta}{1 - x_{jn}}}^{1 + \frac{\eta}{1 - x_{jn}}} |1 - s|^{\alpha_1 - 1} \, ds \\
&\leq C M_n^2 (1 - x_{jn})^{\alpha_1 - 1/2} \left(\frac{\eta}{1 - x_{jn}}\right)^{\alpha_1} \\
&= C M_n^2 \eta^{\alpha_1 - \frac{1}{2}} \left(\frac{\eta}{1 - x_{jn}}\right)^{1/2} \leq C_1 M_n^2 \eta^{\alpha_1 - \frac{1}{2}}.
\end{aligned}
\tag{7.17}
$$

(II) $\eta > \frac{1}{2}\left(1 - x_{jn}\right)$

Here we estimate

$$
J_{32} \leq CM_n^2 \left(1 - x_{jn}\right)^{\alpha_1 - 1/2} \left( \int_0^{3/4} \frac{ds}{\sqrt{|s|}} \right.
$$

$$
\left. + \int_{3/4}^{5/4} |1 - s|^{\alpha_1 - 1} \, ds + \int_{5/4}^{1 + \frac{\eta}{1 - x_{jn}}} |s|^{\alpha_1 - 3/2} \, ds \right)
$$

$$
\leq CM_n^2 \left(1 - x_{jn}\right)^{\alpha_1 - 1/2} \left(1 + \left(\frac{\eta}{1 - x_{jn}}\right)^{\alpha_1 - 1/2}\right)
$$

$$
= CM_n^2 \left(\left(1 - x_{jn}\right)^{\alpha_1 - 1/2} + \eta^{\alpha_1 - 1/2}\right) \leq C_1 M_n^2 \eta^{\alpha_1 - 1/2}. \tag{7.18}
$$

Combining (7.15–7.18), and recalling our assumption that $\alpha \leq \alpha_1 - 1/2$, gives

$$
J_3 = \int_{I_n \cap [x_{jn} - \eta, x_{jn} + \eta]} p_{n,n}^2 (t) \, \bar{Q}_n \left(x_{jn}, t\right) e^{-2nQ_n(t)} dt \leq C_1 M_n^2 \eta^\alpha.
$$

Together with (7.13) and (7.14), and assuming $\alpha \leq \alpha_1 - \frac{1}{2}$, this gives

$$
\int_{I_n} p_{n,n}^2 (t) \, \bar{Q}_n \left(x_{jn}, t\right) e^{-2nQ_n(t)} dt \leq C(\eta) + C_1 M_n^2 \eta^\alpha.
$$

Then (7.10) and (7.12) give

$$
M_n \leq C_0 \left(C(\eta) + C_1 M_n^2 \eta^\alpha\right)^{1/2}.
$$

The crucial thing is that $C_0$ and $C_1$ are independent of $\eta$. Then choosing small enough $\eta$, we obtain

$$
\sup_{n \geq 1} M_n \leq C_2 < \infty.
$$

So we have the upper bound implicit in (7.2) for $m = n$. The lower bounds follow easily: using Theorem 4.2(a),

$$
1 = \int_{I_n} p_{n,m}^2 e^{-2nQ_n} \leq C \int_{-1}^1 p_{n,m}^2 e^{-2nQ_n}
$$

$$
\leq C \left( \sup_{x \in [-1,1]} |p_{n,m}(x)| \, e^{-nQ_n(x)} \left(1 - x^2\right)^{1/4} \right)^2 \int_{-1}^1 \frac{dx}{\sqrt{1 - x^2}}.
$$

(b) We have from Theorem 4.3 with $R = 1$ and

$$P(x) = |p_{n,n}(x)| \left[ \left(1 - x^2\right)^2 + \left(n^{-2/3}\right)^2 \right]^{1/8} \in \mathbb{P}_{n + \frac{1}{2}},$$

$$\sup_{x \in I_n} |p_{n,n}(x)| \left[ \left(1 - x^2\right)^2 + \left(n^{-2/3}\right)^2 \right]^{1/8} e^{-nQ_n(x)}$$

$$\leq C_1 \sup_{|x| \leq 1 - n^{-2/3}} |p_{n,n}(x)| \left[ \left(1 - x^2\right)^2 + \left(n^{-2/3}\right)^2 \right]^{1/8} e^{-nQ_n(x)}$$

$$\leq C_2 \sup_{|x| \leq 1 - n^{-2/3}} |p_{n,n}(x)| \left[ \left(1 - x^2\right)^2 \right]^{1/8} e^{-nQ_n(x)} \leq C_3,$$

by (a). The lower bound follows from (7.2).                                                        □

Now we turn to the bound for general $m$:

*Proof of Theorem 7.1 for $m = m(n)$ Satisfying (7.1).* Recall that $L_{n,r}$ is defined by (3.8). Let

$$r_n = m/n$$

and for $x \in L_{n,r_n}(I_n) =: I_m^\#$, let

$$Q_m^\#(x) = \frac{n}{m} Q_n \left( L_{n,r_n}^{[-1]}(x) \right) = r_n^{-1} Q_n \left( L_{n,r_n}^{[-1]}(x) \right).$$

Observe that

$$\frac{1}{\pi} \int_{-1}^1 \frac{x Q_m^{\#\prime}(x)}{\sqrt{1 - x^2}} \, dx = \frac{1}{\pi} r_n^{-1} \delta_{n,r_n} \int_{-1}^1 \frac{x Q_n' \left( L_{n,r_n}^{[-1]}(x) \right)}{\sqrt{1 - x^2}} \, dx$$

$$= \frac{1}{\pi} r_n^{-1} \int_{a_{-n,r_n}}^{a_{n,r_n}} \frac{L_{n,r_n}(t) \, Q_n'(t)}{\sqrt{1 - L_{n,r_n}^2(t)}} \, dt$$

$$= \frac{1}{\pi} r_n^{-1} \int_{a_{-n,r_n}}^{a_{n,r_n}} \frac{t Q_n'(t)}{\sqrt{(t - a_{-n,r_n})(a_{n,r_n} - t)}} \, dt = 1, \qquad (7.19)$$

by the equilibrium relations (3.1) for $Q_n$. Similarly,

$$\frac{1}{\pi} \int_{-1}^1 \frac{x Q_m^{\#\prime}(x)}{\sqrt{1 - x^2}} \, dx = 0. \qquad (7.20)$$

It follows that $[-1, 1]$ is the Mhaskar-Rakhmanov-Saff interval and the support of the equilibrium measure of total mass 1 for $Q_m^\#$. Also, since $r_n \to 1$ as $n \to \infty$, for each compact subset $\mathcal{K}$ of $(-r^*, r^*)$ and for $m \geq m_0$, where $m_0$ is some large threshold, $Q_m^{\#}$ satisfies a uniform Lipschitz condition of order $\alpha$ on $\mathcal{K} \cap I_m^{\#}$. Next, since

$$r_n = 1 + O\left(n^{-2/3}\right),$$

(3.26) gives

$$\delta_{n,r_n} = 1 + O\left(n^{-2/3}\right) \text{ and } \beta_{n,r_n} = O\left(n^{-2/3}\right).$$

Then we can find a neighborhood $I_0^{\#}$ of $\pm 1$ such that for large enough $n$, $L_{n,r_n}^{[-1]}\left(I_0^{\#}\right) \subset I_0$. Consequently, $x \in I_0^{\#} \Rightarrow L_{n,r_n}^{[-1]}(x) \in I_0$, so $\{Q_m^{\#}\}$ satisfies a Lipschitz condition of order $\alpha_1$ in $I_0^{\#}$. So $\{Q_m^{\#}\}$ lies in the class $\mathcal{Q}$. Next, if $p_m^{\#}$ is the orthonormal polynomial of degree $m$ for $e^{-2mQ_m^{\#}}$, then we claim

$$p_m^{\#}(x) = \sqrt{\delta_{n,r_n}} p_{n,m}\left(L_{n,r_n}^{[-1]}(x)\right).$$

Indeed, defining $p_m^{\#}$ by this last formula,

$$\int_{I_m^{\#}} p_m^{\#}(x)^2 \, e^{-2mQ_m^{\#}(x)} dx$$

$$= \delta_{n,r_n} \int_{L_{n,r_n}(I_n)} p_{n,m}^2\left(L_{n,r_n}^{[-1]}(x)\right) e^{-2nQ_n\left(L_{n,r_n}^{[-1]}(x)\right)} dx$$

$$= \int_{I_n} p_{n,m}^2(t) \, e^{-2nQ_n(t)} dt = 1.$$

The orthogonality of $p_m^{\#}(x)$ to $x^j$, $0 \leq j < m$, is similar. Thus we can apply the special case of Theorem 7.1 to $\{p_m^{\#}\}$, so that for $m = m(n)$ and all $n$,

$$\left|p_m^{\#}(x)\right| e^{-mQ_m^{\#}(x)} \leq C\left[\left|1 - x^2\right| + m^{-2/3}\right]^{-1/4}, \quad \text{for } x \in I_m^{\#}.$$

Setting $x = L_{n,r_n}(t)$ and using $\delta_{n,r_n} \sim 1$, $m \sim n$, we obtain

$$\left|p_{n,m}(t)\right| e^{-nQ_n(t)} \leq C\left[\left|1 - L_{n,r_n}^2(t)\right| + n^{-2/3}\right]^{1/4}, \quad t \in I_n.$$

We need to show that

$$\left|1 - L_{n,r_n}^2(t)\right| + n^{-2/3} \sim \left|1 - t^2\right| + n^{-2/3}, \quad t \in I_n. \tag{7.21}$$

As above, from (3.26), $|\pm 1 - a_{\pm n, r_n}| = O(1 - r_n) = O(n^{-2/3})$, so

$$
\begin{aligned}
\delta_{n,r_n}^2 \left| 1 - L_{n,r_n}^2(t) \right| &= |(t - a_{-n,r_n})(a_{n,r_n} - t)| \\
&= \left| (t + 1 + O(n^{-2/3}))(1 - t + O(n^{-2/3})) \right| \\
&= \left| 1 - t^2 + O(n^{-2/3}) \right|.
\end{aligned}
$$

It follows that if $A$ is large enough, and $|1 - t^2| \geq An^{-2/3}$, then $\left| 1 - L_{n,r_n}^2(t) \right| \sim |1 - t^2|$, and (7.21) follows. On the other hand, if $|1 - t^2| < An^{-2/3}$, this last relation shows that $\left| 1 - L_{n,r_n}^2(t) \right| = O(n^{-2/3})$, and again (7.21) follows. So we have the desired upper bound implicit in (7.3). The lower bound follows as in the case $m(n) = n$. Then (7.2) also follows.                                           $\square$

# Chapter 8
# Markov-Bernstein Inequalities in $L_\infty$

We prove:

**Theorem 8.1.** *Assume* $\{Q_n\} \in \mathcal{Q}$.
*(a) Let* $1 \leq r < r^*$. *Let* $L \geq 0$. *Then for* $n \geq n_0(r)$ *and polynomials* $P$ *of degree* $\leq n + L$,

$$\sup_{x \in [-r,r] \cap I_n} \left| \left( Pe^{-nQ_n} \right)'(x) \max\left\{ 1 - |x|, n^{-2/3} \right\}^{-1/2} \right| \leq Cn \left\| Pe^{-nQ_n} \right\|_{L_\infty(I_n)}. \qquad (8.1)$$

*(b) For* $n \geq 1$ *and polynomials* $P$ *of degree* $\leq n + L$,

$$\left\| P'e^{-nQ_n} \right\|_{L_\infty(I_n)} \leq Cn \left\| Pe^{-nQ_n} \right\|_{L_\infty(I_n)}. \qquad (8.2)$$

*Proof of Theorem 8.1(a).* For the given $r$, we assume that $n$ is so large that $r + n^{-1/2} \leq r^*$. Given $B > 0$, this permits us to apply the Lipschitz condition (1.11) in $\left[ -r - Bn^{-2/3}, r + Bn^{-2/3} \right]$ for $n \geq n_0(B, r)$. Fix $x \in [-r, r] \cap I_n$. Define

$$\hat{Q}_n(t) = Q_n(x) + Q_n'(x)(t - x);$$

$$\hat{W}_n(t) = \exp\left( -\hat{Q}_n(t) \right).$$

Observe that

$$\hat{W}_n^{(j)}(x) = \left( e^{-Q_n} \right)^{(j)}(x), \quad j = 0, 1.$$

© The Author(s) 2018
E. Levin, D.S. Lubinsky, *Bounds and Asymptotics for Orthogonal Polynomials for Varying Weights*, SpringerBriefs in Mathematics,
https://doi.org/10.1007/978-3-319-72947-3_8

Let

$$\rho = \frac{1}{n\sqrt{\max\left\{1 - |x|, n^{-2/3}\right\}}}$$

and $P$ be a polynomial of degree $\leq n$. By Cauchy's estimates for derivatives, and by (4.8) of Lemma 4.4(b), with $T = 1$, $\Omega = n + L$, and as $g_{[-1,1]} \geq 0$,

$$\left|(Pe^{-nQ_n})'(x)\right| = \left|\left(P\hat{W}_n^n\right)'(x)\right|$$

$$\leq \frac{1}{\rho} \max_{|t-x|=\rho} \left|\left(P\hat{W}_n^n\right)(t)\right|$$

$$\leq \frac{1}{\rho} \left\|Pe^{-nQ_n}\right\|_{L_\infty[-1,1]} \max_{|t-x|=\rho} \exp\left\{ \begin{array}{c} -n\left[V^{\sigma Q_n,1}(t) - F_{Q_n,1}\right] + Lg_{[-1,1]}(t) \\ -nQ_n(x) - nQ_n'(x)(\operatorname{Re} t - x) \end{array} \right\}.$$

Setting

$$H_n(x, y) = Q_n(y) - Q_n(x) - Q_n'(x)(y - x),$$

and recalling the definition (3.11) of $U_{n,1}$, we continue this as

$$\leq \frac{1}{\rho} \left\|Pe^{-nQ_n}\right\|_{L_\infty(I_n)} \max_{|t-x|=\rho} \exp\{-n[V^{\sigma Q_n,1}(t) - V^{\sigma Q_n,1}(\operatorname{Re} t)]$$

$$+ Lg_{[-1,1]}(t) + nU_{n,1}(\operatorname{Re} t) + nH_n(x, \operatorname{Re} t)\}. \qquad (8.3)$$

Here $U_{n,1} \leq 0$ and for some $\xi$ between $x$ and $\operatorname{Re} t$,

$$|H_n(\operatorname{Re} t, x)| = \left|\{Q_n'(\xi) - Q_n'(x)\}(\operatorname{Re} t - x)\right|$$

$$\leq C|\xi - x|^\alpha |\operatorname{Re} t - x|$$

$$\leq C\rho^{\alpha+1} \leq \frac{C}{n},$$

provided $\sqrt{\max\left\{1 - |x|, n^{-2/3}\right\}} \geq n^{\frac{1}{1+\alpha}-1}$, by definition of $\rho$. If this last inequality fails, then for $n$ large enough, we are in the interval $I_0$ where $Q_n'$ satisfies a Lipschitz condition of order $\alpha_1 > \frac{1}{2}$, and then instead

$$|H_n(\operatorname{Re} t, x)| \leq C\rho^{\alpha_1+1} \leq C\rho^{3/2} \leq \frac{C}{n},$$

by definition of $\rho$. Also, as $L$ is fixed, and $\rho$ is bounded, $Lg_{[-1,1]}(t) \leq C$ for all $|t - x| = \rho$ and $x \in [-r, r]$. Finally by Lemma 3.5, as $\sigma_{Q_n,1} = \hat{\sigma}_{Q_n,1}$,

$$0 \geq [V^{\sigma_{Q_n,1}}(t) - V^{\sigma_{Q_n,1}}(\text{Re } t)]$$

$$\geq -C |\text{Im } t| \left\{ \sqrt{\max\{1 - |\text{Re } t|, 0\}} + \sqrt{|\text{Im } t|} \right\}$$

$$\geq -C\rho \left\{ \sqrt{\max\{1 - |\text{Re } t|, 0\}} + \sqrt{\rho} \right\}$$

$$\geq -C\rho \left\{ \sqrt{\max\{1 - |x|, 0\}} + \sqrt{\rho} \right\} \geq -\frac{C}{n}.$$

Substituting all these inequalities into (8.3) gives the inequality (8.1).  □

*Proof of Theorem 8.1(b).*  By (a), for $P$ of degree $\leq n$ and $x \in [-1, 1]$,

$$\left| \left( P'(x) e^{-nQ_n(x)} - nQ_n'(x) P(x) e^{-nQ_n(x)} \right) \right| \leq Cn \left\| Pe^{-nQ_n} \right\|_{L_\infty(I_n)}.$$

Since $\{Q_n'\}$ are uniformly bounded in $[-1, 1]$,

$$\left\| P'e^{-nQ_n} \right\|_{L_\infty[-1,1]} \leq Cn \left\| Pe^{-nQ_n} \right\|_{L_\infty(I_n)}.$$

Then Theorem 4.2(a) gives the result.  □

# Chapter 9
# Discretization of Potentials

In this chapter, we discretize potentials, using a method of Totik. We shall develop further properties of these in the next chapter, and then in Chapter 11, will use these to approximate $e^{-2nQ_n}$ by Bernstein-Szegő weights.

**Theorem 9.1.** *For $n \geq 1$, assume that $\sigma_n$ is a nonnegative function on $[-1, 1]$ such that*

$$\int_{-1}^{1} \sigma_n = 1.$$

*Assume further that*

$$\sigma_n(t) = h_n(t) \sqrt{1 - t^2},$$

*where for some $0 < A < B$,*

$$A \leq h_n(t) \leq B, \quad t \in (-1, 1), \tag{9.1}$$

*and for some $C > 0$,*

$$|h_n(t) - h_n(s)| \leq C\omega(|s - t|), \tag{9.2}$$

*where $\omega : [0, \infty) \to [0, \infty)$ is a strictly increasing continuous function with $\omega(0) = 0$, and $\omega(2t) = O(\omega(t))$, $t \geq 0$. The constants $A, B, C$ and $\omega$ are assumed to be independent of $n, s, t$. There exist $n_0, C_0$ and $C_1$ such that if $n \geq n_0$ and $d_n$ is an integer $\geq C_0$, then there exist polynomials $R_n$ of degree $n$ such that*
*(I)*

$$|R_n(u)| \, e^{nV^{\sigma_n}(u)} \geq 1, \quad u \in \mathbb{R}. \tag{9.3}$$

© The Author(s) 2018
E. Levin, D.S. Lubinsky, *Bounds and Asymptotics for Orthogonal Polynomials for Varying Weights*, SpringerBriefs in Mathematics,
https://doi.org/10.1007/978-3-319-72947-3_9

*(II) For $u \in [-1, 1]$ satisfying the inequalities*

$$n (1 - |u|)^{3/2} \geq d_n^6$$  (9.4)

*and*

$$\omega \left( \frac{d_n^4}{n (1 - |u|)^{1/2}} \right) \leq d_n^{-2},$$  (9.5)

*we have*

$$|R_n (u)| \, e^{nV^{\sigma_n}(u)} = e^{d_n \pi + O\left(\frac{1}{d_n}\right)}.$$  (9.6)

*(III) For $u \in \mathbb{C}$,*

$$|R_n (u)| \, e^{nV^{\sigma_n}(u)} \leq e^{C_1 d_n \log n}.$$  (9.7)

To simplify notation, we shall focus on a given $n$, and often omit the subscript $n$. In particular, we let $\sigma = \sigma_n$. Throughout, we assume the hypotheses of Theorem 9.1. For the given $n$, and $\sigma = \sigma_n$, we partition $[-1, 1]$ as

$$-1 = t_0 < t_1 < t_2 < \cdots < t_n = 1,$$

where

$$\int_{t_j}^{t_{j+1}} \sigma = \frac{1}{n}.$$  (9.8)

We let

$$I_j = [t_j, t_{j+1}) \text{ and } |I_j| = t_{j+1} - t_j.$$

We choose the weight point (or center of mass) $\xi_j \in I_j$ by

$$\int_{I_j} (t - \xi_j) \, \sigma (t) \, dt = 0.$$  (9.9)

Define

$$R_n (z) = \prod_{j=0}^{n-1} (z - \xi_j + i d_n |I_j|),$$  (9.10)

where $d_n$ is an integer,

$$d_n \geq 2.$$  (9.11)

Later on, we shall choose $d_n$ to exceed some threshold $C_0$. We also let

$$K = K(n) = d_n^4. \tag{9.12}$$

We let

$$\Gamma_n(u) = \log |R_n(u)| + nV^\sigma(u) \tag{9.13}$$

and observe that

$$\Gamma_n(u) = \sum_{j=0}^{n-1} n \int_{I_j} \log \left| \frac{u - \xi_j + id_n |I_j|}{u - t} \right| \sigma(t)\, dt =: \sum_{j=0}^{n-1} \Gamma_{n,j}(u). \tag{9.14}$$

When we consider $u \in (-1, 1)$, we choose $j_0$ such that

$$u \in I_{j_0}. \tag{9.15}$$

We follow the path in [25, Chapter 7]. First, a lower bound for $\Gamma_{n,j}(u)$:

**Lemma 9.2.** *For* $0 \le j \le n - 1$,

$$\Gamma_{n,j}(u) \ge 0 \text{ for } u \in \mathbb{R}, \tag{9.16}$$

*and hence*

$$\Gamma_n(u) \ge 0 \text{ for } u \in \mathbb{R}.$$

*Proof.* If $u \in I_j$, then for $t \in I_j$,

$$\left| \frac{u - \xi_j + id_n |I_j|}{u - t} \right| \ge \frac{d_n |I_j|}{|I_j|} \ge 1.$$

Thus in this case the integrand in $\Gamma_{n,j}$ is nonnegative, and so $\Gamma_{n,j}(u) \ge 0$. Next, suppose $u$ is to the left of $I_j$. A Taylor expansion of $\log(t - u)$ about $u = \xi_j$ gives, for some $s$ between $\xi_j, t$,

$$\log \left| \frac{u - \xi_j + id_n |I_j|}{u - t} \right| \ge \log \left| \frac{u - \xi_j}{u - t} \right| = \log \left( \frac{\xi_j - u}{t - u} \right)$$

$$= \log(\xi_j - u) - \left[ \log(\xi_j - u) + \frac{t - \xi_j}{\xi_j - u} - \frac{1}{2} \frac{(t - \xi_j)^2}{(s - u)^2} \right]$$

$$\ge -\frac{t - \xi_j}{\xi_j - u},$$

so

$$\Gamma_{n,j}(u) \geq -n \int_{I_j} \frac{t - \xi_j}{\xi_j - u} \sigma(t) \, dt = 0.$$

The case where $u$ is to the right of $I_j$ is similar.                              □

Next, we list some properties of the discretization points. Again, we suppress the dependence of many quantities on $n$, and the constants are uniform in $n$.

**Lemma 9.3.** *(a) For $0 \leq j \leq n-1$,*

$$|I_j| \sim |I_{j+1}| \sim \frac{1}{n\sqrt{1 - t_j^2} + n^{-2/3}}, \tag{9.17}$$

*and*

$$1 - t_j^2 + n^{-2/3} \sim 1 - t_{j+1}^2 + n^{-2/3}. \tag{9.18}$$

*(b)*

$$1 - t_{n-1} \sim n^{-2/3} \sim 1 + t_1. \tag{9.19}$$

*(c) For $2 \leq j \leq n-2$,*

$$\left| \frac{|I_{j+1}|}{|I_j|} - 1 \right| \leq \frac{C}{n\left(1 - t_j^2\right)^{3/2}} + C\omega\left( \frac{1}{n\sqrt{1 - t_j^2}} \right). \tag{9.20}$$

*(d) For $s, t \in (-1, 1)$,*

$$\left| \frac{1}{\sigma(s)} - \frac{1}{\sigma(t)} \right| \leq C \frac{|s - t|}{\min\left\{ \sqrt{1 - s^2}, \sqrt{1 - t^2} \right\}^3} + C \frac{\omega(|s - t|)}{\sqrt{1 - t^2}}, \tag{9.21}$$

*and*

$$|\sigma(s) - \sigma(t)| \leq C \frac{|s - t|}{\max\left\{ \sqrt{1 - s^2}, \sqrt{1 - t^2} \right\}} + C\sqrt{1 - s^2}\,\omega(|s - t|). \tag{9.22}$$

*(e) There exists $C_0$ such that for $d_n \geq C_0$ and $u \in I_{j_0}$ satisfying (9.4),*

$$\left| \frac{t_{j_0 \pm K} - u}{1 - |u|} \right| \leq C \frac{K}{n(1 - |u|)^{3/2}} \leq \frac{1}{2} \tag{9.23}$$

and

$$1 - \left| t_{j_0 \pm \kappa} \right| \sim 1 - |u| . \tag{9.24}$$

*Proof.* (a), (b) From (9.1) and (9.8),

$$\frac{1}{n} \geq A \int_{I_j} \sqrt{1 - t^2}\, dt \geq A \left| I_j \right| \min \left\{ \sqrt{1 - t_j^2}, \sqrt{1 - t_{j+1}^2} \right\} .$$

$$\frac{1}{n} \leq B \int_{I_j} \sqrt{1 - t^2}\, dt \leq B \left| I_j \right| \max \left\{ \sqrt{1 - t_j^2}, \sqrt{1 - t_{j+1}^2} \right\} .$$

If, for example, $t_j \geq 0$, we obtain

$$\frac{1}{nB\sqrt{1 - t_j^2}} \leq \left| I_j \right| \leq \frac{1}{nA\sqrt{1 - t_{j+1}^2}} .$$

Then

$$0 < \sqrt{1 - t_j^2} - \sqrt{1 - t_{j+1}^2}$$

$$= \frac{t_{j+1}^2 - t_j^2}{\sqrt{1 - t_j^2} + \sqrt{1 - t_{j+1}^2}} \leq 2 \frac{\left| I_j \right|}{\sqrt{1 - t_{j+1}^2}}$$

$$\leq \frac{2}{nA\left(1 - t_{j+1}^2\right)} ,$$

so

$$0 < \frac{\sqrt{1 - t_j^2}}{\sqrt{1 - t_{j+1}^2}} - 1 \leq \frac{2}{nA\left(1 - t_{j+1}^2\right)^{3/2}} \leq C$$

if $1 - t_{j+1} \geq \eta n^{-2/3}$, with the last $C$ depending on $\eta$. In this case $\sqrt{1 - t_j^2} \sim \sqrt{1 - t_{j+1}^2}$, and

$$\left| I_j \right| \sim \frac{1}{n\sqrt{1 - t_j^2}} \sim \frac{1}{n\sqrt{1 - t_{j+1}^2}} \sim \left| I_{j+1} \right|$$

and since $1 - t_{j+1}^2 \geq \eta n^{-2/3}$, we also obtain (9.17), (9.18). The case where at least one of $t_j, t_{j+1} \leq 0$ and $1 + t_j \geq \eta n^{-2/3}$ is similar. Finally,

$$\frac{1}{n} = \int_{t_{n-1}}^{1} \sigma(t)\, dt \leq B \int_{t_{n-1}}^{1} \sqrt{1 - t^2}\, dt$$

$$\leq CB\,(1 - t_{n-1})^{3/2}$$

$$\Rightarrow 1 - t_{n-1} \geq Cn^{-2/3}.$$

Similarly $1 - t_{n-1} \leq Cn^{-2/3}$. Thus $1 - t_{n-1} \sim n^{-2/3}$ and similarly, $1 + t_1 \sim n^{-2/3}$, so we have (9.19). If $j = n - 1$, or $j = 1$, we also then trivially have (9.17) and (9.18).

(c) Let us assume that $t_j \geq 0$, so that $I_j, I_{j+1}$ lie in $[0, 1]$. Let us also assume that $j \leq n - 2$. The mean value theorem shows that for some $\xi \in I_{j+1}, \zeta \in I_j$,

$$\left| |I_{j+1}| - |I_j| \right|$$

$$= \left| \frac{1}{\sigma(\xi)} \int_{I_{j+1}} \sigma(t)\, dt - \frac{1}{\sigma(\zeta)} \int_{I_j} \sigma(t)\, dt \right|$$

$$= \frac{1}{n} \left| \frac{1}{h(\xi)\,\sqrt{1 - \xi^2}} - \frac{1}{h(\zeta)\,\sqrt{1 - \zeta^2}} \right|$$

$$= \frac{1}{n} \left| \frac{1}{h(\xi)} \left[ \frac{1}{\sqrt{1 - \xi^2}} - \frac{1}{\sqrt{1 - \zeta^2}} \right] + \frac{1}{\sqrt{1 - \zeta^2}} \left[ \frac{1}{h(\xi)} - \frac{1}{h(\zeta)} \right] \right|$$

$$\leq \frac{C}{n} \left[ \frac{|\xi - \zeta|}{\sqrt{1 - \xi^2}\,\sqrt{1 - \zeta^2}\left(\sqrt{1 - \xi^2} + \sqrt{1 - \zeta^2}\right)} + \frac{\omega(|\xi - \zeta|)}{\sqrt{1 - \zeta^2}} \right]$$

(by (9.1), (9.2))

$$\leq \frac{C}{n} \frac{|I_j|}{\left(1 - t_j^2\right)^{3/2}} + \frac{C}{n} \frac{\omega(|I_j|)}{\left(1 - t_j^2\right)^{1/2}},$$

in view of (9.17), (9.18). Then using (9.17),

$$\left| \frac{|I_{j+1}|}{|I_j|} - 1 \right| \leq \frac{C}{n\left(1 - t_j^2\right)^{3/2}} + C\omega(|I_j|)$$

$$\leq \frac{C}{n\left(1 - t_j^2\right)^{3/2}} + C\omega\left( \frac{1}{n\sqrt{1 - t_j^2}} \right).$$

(d) This was essentially proved in (c).

(e) Let us assume that $u \geq 0$. By the Mean Value Theorem, for some $\xi$ between $u$ and $\frac{1}{2}\left(u + t_{j_0+K}\right)$,

$$\frac{1}{2}\left(t_{j_0+K} - u\right) = \frac{1}{\sigma(\xi)} \int_u^{\frac{1}{2}\left(t_{j_0+K}+u\right)} \sigma(t)\, dt$$

$$\leq \frac{C}{\sqrt{1 - \left[\frac{1}{2}\left(t_{j_0+K} + u\right)\right]^2}} \int_{t_{j_0}}^{t_{j_0+K}} \sigma(t)\, dt$$

$$\leq \frac{C}{\sqrt{1 - \left[\frac{1}{2}\left(t_{j_0+K} + u\right)\right]}} \frac{K}{n}$$

$$\leq \frac{C}{\sqrt{1 - u}} \frac{K}{n}.$$

Then from (9.4) and (9.12),

$$\frac{t_{j_0+K} - u}{1 - u} \leq C \frac{K}{n\left(1 - u\right)^{3/2}} \leq C \frac{1}{d_n^2} \leq \frac{1}{2},$$

provided $d_n \geq C_0 = \sqrt{2C}$. Also then

$$0 < \frac{1 - u}{1 - t_{j_0+K}} - 1 = \frac{t_{j_0+K} - u}{1 - u} \leq \frac{1}{2},$$

so (9.24) also follows. The remaining cases are handled similarly.  $\square$

Next, we estimate the "tail" terms:

**Lemma 9.4.** *Let $u \in I_{j_0}$ and*

$$\sum_1 = \sum_{j=0}^{j_0-K-1} \Gamma_{n,j}(u); \quad \sum_3 = \sum_{j=j_0+K}^{n-1} \Gamma_{n,j}(u). \tag{9.25}$$

*Then*

$$\sum_1 + \sum_3 \leq \frac{C}{d_n^2}. \tag{9.26}$$

*Proof.* Note first that

$$|j - j_0| \geq 2 \Rightarrow \mathrm{dist}\left(u, I_j\right) \geq C\left|I_j\right|,$$

as follows from $\left|I_j\right| \sim \left|I_{j+1}\right|$. Next if $t \in I_j$, we write

$$\frac{u - \xi_j + id_n\left|I_j\right|}{u - t} = 1 + \frac{t - \xi_j}{u - t} + i\frac{d_n\left|I_j\right|}{u - t} = 1 + S + iT,$$

say. Here

$$\left|T/S\right| = d_n\left|I_j\right|/\left|t - \xi_j\right| \geq d_n\left|I_j\right|/\left|I_j\right| \geq 2.$$

By Lemma 7.18 in [25, p. 196], if $|T| \geq |S|$ or $S \geq -\frac{1}{2}$,

$$\left|\log\left|1 + S + iT\right| - S\right| \leq C\left(S^2 + T^2\right), \tag{9.27}$$

where $C$ is an absolute constant, that is

$$\left|\log\left|\frac{u - \xi_j + id_n\left|I_j\right|}{u - t}\right| - \frac{t - \xi_j}{u - t}\right| \leq C\left(\left(\frac{t - \xi_j}{u - t}\right)^2 + \left(\frac{d_n\left|I_j\right|}{u - t}\right)^2\right)$$

$$\leq C\left(\frac{d_n\left|I_j\right|}{u - t}\right)^2.$$

Also

$$\frac{t - \xi_j}{u - t} = \frac{t - \xi_j}{u - \xi_j} + \frac{\left(t - \xi_j\right)^2}{\left(u - t\right)\left(u - \xi_j\right)} = \frac{t - \xi_j}{u - \xi_j} + O\left(\frac{\left(t - \xi_j\right)^2}{\left(u - t\right)^2}\right).$$

Then

$$\log\left|\frac{u - \xi_j + id_n\left|I_j\right|}{u - t}\right| = \frac{t - \xi_j}{u - \xi_j} + O\left(\frac{d_n\left|I_j\right|}{u - t}\right)^2,$$

so using the definition of $\xi_j$,

$$\Gamma_{n,j}(u) = n\int_{I_j}\left\{\frac{t - \xi_j}{u - \xi_j} + O\left(\frac{d_n\left|I_j\right|}{u - t}\right)^2\right\}\sigma(t)\,dt$$

$$= 0 + n\left(d_n\left|I_j\right|\right)^2 O\left(\int_{I_j}\frac{\sqrt{1 - t^2}}{\left(u - t\right)^2}dt\right)$$

$$= O\left(\frac{d_n^2}{n}\int_{I_j}\frac{1}{\left(u - t\right)^2}\frac{dt}{\sqrt{1 - t^2}}\right),$$

recall (9.17–18). Then

$$\sum_1 + \sum_3 \leq C \frac{d_n^2}{n} \left\{ \int_{-1}^{t_{j_0}-K} + \int_{t_{j_0}+K}^{1} \right\} \frac{1}{(u-t)^2} \frac{dt}{\sqrt{1-t^2}}.$$

Suppose, for example, $u \geq 0$. Then we continue this, using the substitution $(1-t) = s(1-u)$ as

$$\leq C \frac{d_n^2}{n} \left\{ \int_{-1}^{-\frac{1}{2}} \frac{dt}{\sqrt{1-t^2}} + \left[ \int_{-\frac{1}{2}}^{t_{j_0}-K} + \int_{t_{j_0}+K}^{1} \right] \frac{1}{(u-t)^2} \frac{dt}{\sqrt{1-t}} \right\}$$

$$\leq C \frac{d_n^2}{n} \left\{ 1 + (1-u)^{-3/2} \left[ \int_{\frac{1-t_{j_0}-K}{1-u}}^{\frac{3}{2(1-u)}} + \int_{0}^{\frac{1-t_{j_0}+K}{1-u}} \right] \frac{1}{(1-s)^2 \sqrt{s}} ds \right\}$$

$$\leq C \frac{d_n^2}{n} \left\{ 1 + (1-u)^{-3/2} \left[ 1 + \left( \frac{1-t_{j_0}-K}{1-u} - 1 \right)^{-1} + \left( 1 - \frac{1-t_{j_0}+K}{1-u} \right)^{-1} \right] \right\}$$

$$\leq C \frac{d_n^2}{n} \left\{ (1-u)^{-3/2} \left[ 1 + \frac{1-u}{u-t_{j_0}-K} + \frac{1-u}{t_{j_0}+K-u} \right] \right\}.$$

Next observe that

$$u - t_{j_0-K} \geq \int_{t_{j_0}-K}^{t_{j_0}} dt \geq C \int_{t_{j_0}-K}^{t_{j_0}} \frac{\sigma(t)}{\sqrt{1-t^2}} dt \geq \frac{CK}{n\sqrt{1-u}}, \qquad (9.28)$$

by (9.24), with a similar inequality for $t_{j_0+K} - u$. Thus

$$\sum_1 + \sum_3 \leq C \frac{d_n^2}{n(1-u)^{3/2}} + C \frac{d_n^2}{K} \leq \frac{C}{d_n^2},$$

recall (9.4) and (9.12).                                                                                    □

Now we handle the central terms:

**Lemma 9.5.** *Let*

$$\sum_2 = \sum_{j=j_0-K}^{j_0+K-1} \Gamma_{n,j}(u). \qquad (9.29)$$

*Then*

$$\sum_2 = d_n \pi + O\left(\frac{1}{d_n}\right). \qquad (9.30)$$

*Proof.* We write (as in the proof of Lemma 7.21 in [25, p. 203]),

$$\Gamma_{n,j}(u) = n \int_{I_j} \log \left| \frac{u - \xi_j + id_n |I_j|}{u - t + id_n/(n\sigma(u))} \right| \sigma(t)\, dt$$

$$+ n \int_{I_j} \log \left| \frac{u - t + id_n/(n\sigma(u))}{u - t} \right| \sigma(t)\, dt$$

$$=: A_{j,1} + A_{j,2}.$$

We shall show that

$$\sum_{j=j_0-K}^{j_0+K-1} |A_{j,1}| \leq \frac{C}{d_n}; \tag{9.31}$$

$$\sum_{j=j_0-K}^{j_0+K-1} A_{j,2} = d_n \pi + O\left(\frac{1}{d_n}\right). \tag{9.32}$$

These together give (9.30).                                                    □

*Proof of* (9.31). We assume that $t_{j_0} \geq 0$. We write for $j_0 - K \leq j \leq j_0 + K - 1$,

$$A_{j,1} = n \int_{I_j} \log \left| 1 + \rho_j(t) \right| \sigma(t)\, dt,$$

where

$$\rho_j(t) = \frac{t - \xi_j + id_n \left[ |I_j| - \frac{1}{n\sigma(u)} \right]}{u - t + id_n/(n\sigma(u))}.$$

Here for some $\xi \in I_j$, (9.21) gives

$$\left| |I_j| - \frac{1}{n\sigma(u)} \right| = \frac{1}{n} \left| \frac{1}{\sigma(\xi)} - \frac{1}{\sigma(u)} \right|$$

$$\leq \frac{C}{n} \left[ \frac{|\xi - u|}{\min\left\{ \sqrt{1 - \xi^2}, \sqrt{1 - u^2} \right\}^3} + \frac{\omega(|\xi - u|)}{\sqrt{1 - u^2}} \right]$$

$$\leq \frac{C}{n} \left[ \frac{t_{j_0+K} - t_{j_0-K}}{\left( \sqrt{1 - t_{j_0+K}} \right)^3} + \frac{\omega\left( t_{j_0+K} - t_{j_0-K} \right)}{\sqrt{1 - t_{j_0+K}}} \right]$$

$$\leq \frac{C}{n} \left[ \frac{K}{n(1-u)^2} + \frac{1}{\sqrt{1-u}} \omega\left( \frac{K}{n\sqrt{1-u}} \right) \right]$$

$$\leq \frac{C}{d_n^2} |I_j|, \tag{9.33}$$

by (9.4), (9.5), (9.17), and (9.23). Then

$$|\rho_j(t)| \le C\frac{|I_j|}{\sqrt{(u-t)^2 + (d_n|I_j|)^2}} \le \frac{C}{d_n} \le \frac{1}{2}, \qquad (9.34)$$

for large enough $n$, if $d_n \ge C_0$ for some fixed $C_0$. Then using the inequality (9.27) (which is applicable as Re $\rho_j(t) \ge -\frac{1}{2}$)

$$\big|\log|1 + \rho_j(t)| - \text{Re } \rho_j(t)\big| \le C|\rho_j(t)|^2. \qquad (9.35)$$

Here

$$
\text{Re } \rho_j(t) = \frac{(t-\xi_j)(u-t)}{(u-t)^2 + (d_n/(n\sigma(u)))^2} + \frac{d_n^2\left[|I_j| - \frac{1}{n\sigma(u)}\right]\frac{1}{n\sigma(u)}}{(u-t)^2 + (d_n/(n\sigma(u)))^2}
$$

$$
= (t-\xi_j)\,Q(u-t, d_n/(n\sigma(u))) + \frac{d_n^2\left[|I_j| - \frac{1}{n\sigma(u)}\right]\frac{1}{n\sigma(u)}}{(u-t)^2 + (d_n/(n\sigma(u)))^2}, \qquad (9.36)
$$

where $Q$ is the conjugate Poisson kernel, given by

$$Q(r,s) = \frac{r}{r^2 + s^2}.$$

As

$$\left|\frac{\partial Q(r,s)}{\partial r}\right| = \left|\frac{r^2 - s^2}{(r^2 + s^2)^2}\right| \le \frac{1}{r^2 + s^2},$$

so

$$\big|Q(u-t, d_n/(n\sigma(u))) - Q(u-\xi_j, d_n/(n\sigma(u)))\big|$$

$$\le \frac{C|t-\xi_j|}{(u-t)^2 + (d_n/(n\sigma(u)))^2}$$

$$\le C\frac{|I_j|}{(u-t)^2 + (d_n/(n\sigma(u)))^2},$$

so using also (9.33),

$$\big|\text{Re } \rho_j(t) - (t-\xi_j)\,Q(u-\xi_j, d_n/(n\sigma(u)))\big| \le C\frac{|I_j|^2}{(u-t)^2 + (d_n/(n\sigma(u)))^2}.$$

By definition of $A_{j,1}$, we have

$$\left|A_{j,1}\right| \leq n \int_{I_j} \left|\log\left|1 + \rho_j(t)\right| - \operatorname{Re} \rho_j(t)\right| \sigma(t)\,dt$$

$$+ n \int_{I_j} \left|\operatorname{Re} \rho_j(t) - (t - \xi_j)\, Q\left(u - \xi_j, d_n/\left(n\sigma(u)\right)\right)\right| \sigma(t)\,dt$$

$$+ n \left|Q\left(u - \xi_j, d_n/\left(n\sigma(u)\right)\right) \int_{I_j} (t - \xi_j)\,\sigma(t)\,dt\right|.$$

Using the bound above and (9.34), (9.35) gives

$$\left|A_{j,1}\right| \leq Cn \int_{I_j} \frac{\left|I_j\right|^2}{(u - t)^2 + \left(d_n/\left(n\sigma(u)\right)\right)^2}\,\sigma(t)\,dt + 0$$

$$\leq \frac{C}{n\sigma(u)} \int_{I_j} \frac{dt}{(u - t)^2 + \left(d_n/\left(n\sigma(u)\right)\right)^2},$$

recall (9.24) of Lemma 9.3(e). Adding over $j$,

$$\sum_{j=j_0-K}^{j_0+K-1} \left|A_{j,1}\right| \leq \frac{C}{n\sigma(u)} \int_{-\infty}^{\infty} \frac{dt}{(u - t)^2 + \left(d_n/\left(n\sigma(u)\right)\right)^2} = \frac{C}{d_n}\pi.$$

$\square$

*Proof of (9.32).*

$$\sum_{j=j_0-K}^{j_0+K-1} A_{j,2} = \frac{n}{2} \int_{t_{j_0-K}}^{t_{j_0+K}} \log\left(1 + \left(\frac{d_n}{n\sigma(u)(u - t)}\right)^2\right)\sigma(t)\,dt$$

$$= \frac{n\sigma(u)}{2} \int_{t_{j_0-K}}^{t_{j_0+K}} \log\left(1 + \left(\frac{d_n}{n\sigma(u)(u - t)}\right)^2\right)dt$$

$$+ \frac{n}{2} \int_{t_{j_0-K}}^{t_{j_0+K}} \log\left(1 + \left(\frac{d_n}{n\sigma(u)(u - t)}\right)^2\right)[\sigma(t) - \sigma(u)]\,dt$$

$$=: J_1 + J_2.$$

Here the substitution $\frac{d_n}{n\sigma(u)(u-t)} = \frac{1}{s}$ shows that

$$J_1 = \frac{d_n}{2} \int_{n\sigma(u)\left(\frac{u-t_{j_0}+K}{d_n}\right)}^{n\sigma(u)\left(\frac{u-t_{j_0}-K}{d_n}\right)} \log\left(1 + s^{-2}\right) ds$$

$$= \frac{d_n}{2}\left[\int_{-\infty}^{\infty} \log\left(1 + s^{-2}\right) ds + O\left(\left(n\sigma(u)\left|\frac{u - t_{j_0}\pm K}{d_n}\right|\right)^{-1}\right)\right].$$

Here [16, p. 560, no. 4.295.3] (or an integration by parts) gives

$$\int_{-\infty}^{\infty} \log\left(1 + s^{-2}\right) ds = \int_{-\infty}^{\infty} \log\left(1 + v^2\right) \frac{1}{v^2} dv = 2\pi$$

while from (9.17) and (9.24),

$$n\sigma(u)\left|\frac{u - t_{j_0}\pm K}{d_n}\right| \sim \frac{n\sqrt{1 - |u|}}{d_n}\left|\int_u^{t_{j_0}\pm K} dt\right|$$

$$\sim \frac{n}{d_n}\left|\int_u^{t_{j_0}\pm K} \sigma(t)\, dt\right| \sim \frac{K}{d_n},$$

so

$$J_1 = d_n\pi + O\left(\frac{d_n^2}{K}\right) = d_n\pi + O\left(\frac{1}{d_n^2}\right).$$

Next, for $t \in \left[t_{j_0-K}, t_{j_0+K}\right]$, (9.22) gives

$$|\sigma(t) - \sigma(u)|$$

$$\leq C\sigma(u)\left[\frac{|t - u|}{1 - u^2} + \omega\left(|u - t|\right)\right]$$

$$\leq C\sigma(u)\left[\frac{t_{j_0+K} - t_{j_0-K}}{1 - u^2} + \omega\left(t_{j_0+K} - t_{j_0-K}\right)\right]$$

$$\leq C\sigma(u)\left\{\frac{CK}{n\left(1 - u^2\right)^{3/2}} + \omega\left(\frac{K}{n\left(1 - u^2\right)^{1/2}}\right)\right\}$$

$$\leq C\sigma(u)/d_n^2,$$

by (9.23), (9.4), (9.5), and (9.12). Then using our later estimates for $J_1$,

$$J_2 \leq \frac{Cn}{d_n^2} \sigma(u) \int_{t_{j_0}-\kappa}^{t_{j_0}+\kappa} \log\left(1 + \left(\frac{d_n}{n\sigma(u)(u-t)}\right)^2\right) dt$$

$$\leq \frac{C}{d_n^2} J_1 \leq \frac{C}{d_n}.$$

So we have (9.32) and hence also (9.30).                                          $\square$

*Proof of Theorem* 9.1.   From (9.13),

$$\log\left[|R_n(u)| e^{nV^\sigma(u)}\right] = \Gamma_n(u).$$

(I) By Lemma 9.2, $\Gamma_n(u) \geq 0$, which is equivalent to (9.3).
(II) From Lemma 9.4 and 9.5, for $u$ satisfying (9.4) and (9.5), we have

$$\Gamma_n(u) = \sum\nolimits_1 + \sum\nolimits_2 + \sum\nolimits_3 = d_n\pi + O\left(\frac{1}{d_n}\right).$$

This gives (9.6).
(III) Let first $u \in [-1, 1]$. We assume that $u \in I_{j_0}$ and split

$$\Gamma_n(u) = \sum_{j=0}^{j_0-2} + \sum_{j=j_0-1}^{j_0+1} + \sum_{j=j_0+2}^{n-1} =: S_1 + S_2 + S_3.$$

Here the central term $S_2$ contains at most 3 terms (and 2 if $j_0 = 0$ or $n-1$). Now for each $j_0 - 1 \leq j \leq j_0 + 1$, we have

$$\Gamma_{n,j}(u) \leq n \int_{I_j} \log \frac{Cd_n |I_j|}{|u-t|} \sigma(t)\, dt$$

$$= \log(Cd_n) + n \int_{I_j} \log \frac{|I_j|}{|u-t|} \sigma(t)\, dt$$

$$\leq \log(Cd_n) + Cn \int_{I_j} \log \frac{|I_j|}{|u-t|} \sqrt{1-t^2}\, dt$$

$$\leq \log(Cd_n) + C \int_{I_j} \log \frac{|I_j|}{|u-t|} \frac{dt}{|I_j|} \quad \text{(by (9.17-18))}$$

$$\leq \log(Cd_n) + C \int_{-1}^{1} \log \frac{1}{|s|} ds.$$

Thus

$$S_2 \leq \log(Cd_n) + C.$$

Next, consider $S_3$. For $j_0 + 1 \leq j \leq n - 1$, set

$$A_j = \left|I_{j_0+1}\right| + \left|I_{j_0+2}\right| + \cdots + \left|I_j\right| = t_{j+1} - t_{j_0+1}.$$

Then for $j \geq j_0 + 2$ and $t \in I_j$,

$$t - u \geq t_j - t_{j_0+1} = A_{j-1}.$$

Then for $t \in I_j$,

$$\log\left|\frac{u - \xi_j + id_n\left|I_j\right|}{u - t}\right| \leq \log\left|\frac{|u - t| + (d_n + 1)\left|I_j\right|}{|u - t|}\right|$$

$$\leq \log\left(1 + \frac{(d_n + 1)\left|I_j\right|}{A_{j-1}}\right)$$

$$= \log\left(\frac{A_j + d_n\left|I_j\right|}{A_{j-1}}\right)$$

$$= \log\left(\frac{A_j}{A_{j-1}}\right) + \log\left(1 + \frac{d_n\left|I_j\right|}{A_j}\right)$$

$$\leq \log\left(\frac{A_j}{A_{j-1}}\right) + \frac{d_n\left|I_j\right|}{A_j}.$$

Then as $n \int_{I_j} \sigma = 1$,

$$S_3 \leq \sum_{j=j_0+2}^{n-1} \log\left(\frac{A_j}{A_{j-1}}\right) + d_n \sum_{j=j_0+2}^{n-1} \frac{\left|I_j\right|}{A_j}$$

$$\leq \log\frac{A_{n-1}}{A_{j_0+1}} + d_n \sum_{j=j_0+2}^{n-1} \frac{A_j - A_{j-1}}{A_j}$$

$$\leq \log\frac{2}{A_{j_0+1}} + d_n \int_{A_{j_0+1}}^{2} \frac{dt}{t}$$

$$= \log\frac{2}{\left|I_{j_0+1}\right|} + d_n \int_{\left|I_{j_0+1}\right|}^{2} \frac{dt}{t}$$

$$\leq C\log n + Cd_n \log n.$$

A similar estimate holds for $S_1$. Thus for $u \in [-1, 1]$,

$$\Gamma_n(u) \leq \log d_n + C + C d_n \log n.$$

Finally for $u$ outside $[-1, 1]$, we can use the fact that $\Gamma_n(u)$ is subharmonic in $\mathbb{C} \backslash [-1, 1]$, and has limit 0 at $\infty$, so is subharmonic and bounded above in $\overline{\mathbb{C}} \backslash [-1, 1]$. The maximum principle for subharmonic functions gives for all $u \in \mathbb{C}$,

$$\Gamma_n(u) \leq C d_n \log n.$$

$\square$

# Chapter 10
# Derivatives of Discretized Polynomials

In this chapter, we estimate the derivative of the discretized potentials from the last chapter. We prove

**Theorem 10.1.** *Assume the hypotheses and notation of Theorem 9.1. In addition assume that $\int_0^1 \frac{\omega(t)}{t}dt < \infty$. The polynomials $R_n$ of Theorem 9.1 satisfy*

$$\frac{1}{n}\left|\frac{d}{du}\log\left\{|R_n(u)|\,e^{nV^{\sigma_n}(u)}\right\}\right| \le \frac{C}{d_n}$$

*for $u \in [-1, 1]$ satisfying (9.4) and (9.5), and the additional restriction*

$$\int_0^{d_n^2/(n\sqrt{1-|u|})} \frac{\omega(s)}{s}ds \le \frac{1}{d_n}. \tag{10.1}$$

*The constant $C$ is independent of $n, u$, and the particular $\sigma_n$, depending only on $A, B, \omega$.*

As in the previous chapter, we omit many subscripts involving $n$ and use $\sigma = \sigma_n$. Recall that $\Gamma_n(u) = \log\left\{|R_n(u)|\,e^{nV^{\sigma_n}(u)}\right\}$ satisfies (9.14). In view of (9.8), we can reformulate (9.14) as

$$\Gamma_n(u) = \sum_{j=0}^{n-1}\log\left|u - \xi_j + id_n\,|I_j|\right| - n\int_{-1}^{1}\log|u - t|\,\sigma(t)\,dt.$$

© The Author(s) 2018
E. Levin, D.S. Lubinsky, *Bounds and Asymptotics for Orthogonal Polynomials for Varying Weights*, SpringerBriefs in Mathematics,
https://doi.org/10.1007/978-3-319-72947-3_10

Then as $\sigma$ satisfies a uniform Dini condition, we can differentiate to obtain a principal value integral

$$\Gamma_n'(u) = \sum_{j=0}^{n-1} \frac{u - \xi_j}{\left(u - \xi_j\right)^2 + \left(d_n \left|I_j\right|\right)^2} - nPV \int_{-1}^{1} \frac{1}{u - t} \sigma(t) \, dt, \tag{10.2}$$

and as at (9.14),

$$\Gamma_n'(u) = \sum_{j=0}^{n-1} n \int_{I_j} \left[ \frac{u - \xi_j}{\left(u - \xi_j\right)^2 + \left(d_n \left|I_j\right|\right)^2} - \frac{1}{u - t} \right] \sigma(t) \, dt$$

$$= \sum_{j=0}^{n-1} \Gamma_{n,j}^{(1)}(u). \tag{10.3}$$

As in the previous chapter, we fix $u$, and assume $u \in I_{j_0}$. We set

$$L = d_n^2 \tag{10.4}$$

and for $0 \le j \le n - 1$,

$$\tau_j = \left(u - \xi_j\right)^2 + \left(d_n \left|I_j\right|\right)^2. \tag{10.5}$$

Throughout we assume (9.4), (9.5), and (10.1). We start with some technical estimates:

**Lemma 10.2.** *(a) For $|j - j_0| \le L$,*

$$\left|I_j\right| \sim \left|I_{j_0}\right| \sim \frac{1}{n\sqrt{1 - u^2}} \tag{10.6}$$

*and*

$$1 - t_j^2 \sim 1 - u^2. \tag{10.7}$$

*(b) For $2 \le |j - j_0| \le L$,*

$$\left|u - \xi_j\right| \sim |j - j_0| \left|I_{j_0}\right|. \tag{10.8}$$

*(c) For* $1 \leq k \leq L$,

$$\left| (u - \xi_{j_0-k}) + (u - \xi_{j_0+k}) \right|$$

$$\leq C \left| I_{j_0} \right| \left\{ 1 + \frac{k^2}{n(1-u)^{3/2}} + k\omega \left( \frac{k}{n\sqrt{1-u}} \right) \right\}$$

$$=: C\rho_k, \tag{10.9}$$

*say.*

*(d) For* $1 \leq k \leq L$,

$$k \left| \left| I_{j_0+k} \right| - \left| I_{j_0-k} \right| \right| \leq C\rho_k. \tag{10.10}$$

*Proof.* We do this for $u \geq 0$.

(a) From (9.17),

$$\frac{\left| I_{j_0} \right|}{\left| I_j \right|} \sim \sqrt{\frac{1 - t_j^2 + n^{-2/3}}{1 - t_{j_0}^2 + n^{-2/3}}} = \sqrt{1 + \frac{t_{j_0}^2 - t_j^2}{1 - t_{j_0}^2 + n^{-2/3}}}.$$

Here as $|j - j_0| \leq L = d_n^2 \ll K = d_n^4$, we see from (9.23) and then (9.4) that

$$\left| \frac{t_{j_0}^2 - t_j^2}{1 - t_{j_0}^2 + n^{-2/3}} \right| \leq C \left| \frac{t_{j_0} - t_{j_0 \pm K}}{1 - |t_{j_0}|} \right|$$

$$\leq \frac{CK}{n(1-|u|)^{3/2}} \leq \frac{C}{d_n^2} < \frac{1}{2},$$

so $|I_j| \sim |I_{j_0}|$. Then (9.17) gives the last $\sim$ in (10.6). Finally,

$$\frac{1 - t_j}{1 - u} = 1 + \frac{u - t_j}{1 - u},$$

and by (9.23) again,

$$\left| \frac{u - t_j}{1 - u} \right| \leq \left| \frac{t_{j_0} - t_{j_0 \pm K}}{1 - u} \right| \leq \frac{1}{2}.$$

(b) This follows easily from (a).

(c) In view of (a),

$$\left|(u - \xi_{j_0-k}) + (u - \xi_{j_0+k})\right|$$
$$\leq C\left|I_{j_0}\right| + \left|(t_{j_0} - t_{j_0-k}) - (t_{j_0+k} - t_{j_0})\right|$$
$$= C\left|I_{j_0}\right| + \left|\frac{1}{\sigma(\xi)}\int_{t_{j_0-k}}^{t_{j_0}} \sigma(t)\,dt - \frac{1}{\sigma(\zeta)}\int_{t_{j_0}}^{t_{j_0+k}} \sigma(t)\,dt\right|$$
$$= C\left|I_{j_0}\right| + \frac{k}{n}\left|\frac{1}{\sigma(\xi)} - \frac{1}{\sigma(\zeta)}\right|,$$

for some $\xi, \zeta \in [t_{j_0-k}, t_{j_0+k}]$, by the mean value theorem. Using (9.21), we continue this as

$$\leq C\left|I_{j_0}\right| + C\frac{k}{n}\left\{\frac{t_{j_0+k+1} - t_{j_0-k}}{\min\left\{\sqrt{1 - t_{j_0+k+1}^2}, \sqrt{1 - t_{j_0-k}^2}\right\}^3} + \frac{\omega\left(t_{j_0+k+1} - t_{j_0-k}\right)}{\sqrt{1 - t_{j_0+k+1}^2}}\right\}$$

$$\leq C\left|I_{j_0}\right|\left\{1 + \frac{k^2}{n\left(\sqrt{1-u}\right)^3}\right\} + C\left|I_{j_0}\right|k\omega\left(\frac{k}{n\sqrt{1-u}}\right) = C\rho_k,$$

as at (10.9).

(d) For some $\xi \in I_{j_0+k}$ and $\zeta \in I_{j_0-k}$

$$k\left|\left|I_{j_0+k}\right| - \left|I_{j_0-k}\right|\right| = k\left|\frac{1}{\sigma(\xi)}\int_{I_{j_0+k}} \sigma(t)\,dt - \frac{1}{\sigma(\zeta)}\int_{I_{j_0-k}} \sigma(t)\,dt\right|$$

$$= \frac{k}{n}\left|\frac{1}{\sigma(\xi)} - \frac{1}{\sigma(\zeta)}\right| \leq C\rho_k,$$

as in (c).                                                                              □

First we estimate tail terms:

**Lemma 10.3.**

$$\sum_{j:|j-j_0|>L} \left|\Gamma_{n,j}^{(1)}(u)\right| \leq C\frac{n}{d_n^2}. \tag{10.11}$$

*Proof.* Let $|j - j_0| > L$ and $t \in I_j$, while as usual, $u \in I_{j_0}$. Then if $\text{dist}(u, I_j)$ denotes the distance from $u$ to $I_j$,

$$\frac{u - \xi_j}{(u - \xi_j)^2 + (d_n |I_j|)^2} - \frac{1}{u - t}$$

$$= \left( \frac{u - \xi_j}{(u - \xi_j)^2 + (d_n |I_j|)^2} - \frac{1}{u - \xi_j} \right) + \left( \frac{1}{u - \xi_j} - \frac{1}{u - t} \right)$$

$$= \frac{-(d_n |I_j|)^2}{\left[ (u - \xi_j)^2 + (d_n |I_j|)^2 \right](u - \xi_j)} + \frac{\xi_j - t}{(u - \xi_j)(u - t)}$$

$$= \frac{-(d_n |I_j|)^2}{\left[ (u - \xi_j)^2 + (d_n |I_j|)^2 \right](u - \xi_j)} + \frac{\xi_j - t}{u - \xi_j} \left\{ \frac{1}{u - \xi_j} + \frac{t - \xi_j}{(u - t)(u - \xi_j)} \right\}$$

$$= O\left( \frac{(d_n |I_j|)^2}{\text{dist}(u, I_j)^3} \right) + \frac{\xi_j - t}{(u - \xi_j)^2} + O\left( \frac{|I_j|^2}{\text{dist}(u, I_j)^3} \right)$$

$$= O\left( \frac{(d_n |I_j|)^2}{\text{dist}(u, I_j)^3} \right) + \frac{\xi_j - t}{(u - \xi_j)^2}.$$

Integrating over $I_j$, and using the weight point property (9.9),

$$\Gamma_{n,j}^{(1)}(u) = n \int_{I_j} O\left( \frac{(d_n |I_j|)^2}{\text{dist}(u, I_j)^3} \right) \sigma(t)\, dt + 0$$

$$= O\left( \frac{(d_n |I_j|)^2}{\text{dist}(u, I_j)^3} \right)$$

$$= O\left( \frac{d_n^2}{n \sqrt{1 - t_j^2} + n^{-2/3}} \frac{|I_j|}{\text{dist}(u, I_j)^3} \right),$$

by (9.17) and (9.8). Adding, and estimating in an obvious way gives

$$\sum_{j:|j - j_0| > L} \left| \Gamma_{n,j}^{(1)}(u) \right| \leq C \frac{d_n^2}{n} \left\{ \int_{-1}^{t_{j_0} - L} + \int_{t_{j_0} + L}^{1} \right\} \frac{dt}{|u - t|^3 \sqrt{1 - t^2}}.$$

Let us assume that $u \geq 0$. Then with the substitution $(1 - t) = s(1 - u)$, we see that we can continue this as

$$
\leq C \frac{d_n^2}{n(1-u)^{5/2}} \left\{ \int_{\frac{1-t_{j_0}-L}{1-u}}^{\frac{2}{1-u}} + \int_0^{\frac{1-t_{j_0}+L}{1-u}} \right\} \frac{ds}{|1-s|^3 \sqrt{|s|}}
$$

$$
\leq C \frac{d_n^2}{n(1-u)^{5/2}} \left\{ 1 + \left( \frac{1-t_{j_0}-L}{1-u} - 1 \right)^{-2} + \left( 1 - \frac{1-t_{j_0}+L}{1-u} \right)^{-2} \right\}
$$

$$
\leq C \frac{d_n^2}{n(1-u)^{5/2}} \left\{ 1 + \left( \frac{u-t_{j_0}-L}{1-u} \right)^{-2} + \left( \frac{t_{j_0}+L-u}{1-u} \right)^{-2} \right\}
$$

$$
\leq C \frac{d_n^2}{n(1-u)^{5/2}} \left\{ 1 + \left( \frac{1-u}{L |I_{j_0}|} \right)^2 \right\} \quad \text{(by (10.6), (10.8))}
$$

$$
\leq C \frac{d_n^2}{n(1-u)^{5/2}} \left\{ 1 + \left( \frac{n(1-u)^{3/2}}{L} \right)^2 \right\}
$$

$$
= C \frac{d_n^2}{n(1-u)^{5/2}} + C \frac{n}{d_n^2} (1-u)^{1/2} \leq C \frac{n}{d_n^2}.
$$

We used (9.4) to bound the first term.                                        □

Next we turn to the central terms, which are the most difficult. We separately estimate the terms in the sum and the central integral.

**Lemma 10.4.**

$$
\sum_1 = \sum_{j=j_0-L}^{j_0+L} \frac{u - \xi_j}{(u - \xi_j)^2 + (d_n |I_j|)^2} \leq C \frac{n}{d_n}. \tag{10.12}
$$

*Proof.* We split off the term for $j = j_0$, so

$$
\sum_1 = \frac{u - \xi_{j_0}}{(u - \xi_{j_0})^2 + (d_n |I_{j_0}|)^2} + \sum_{1,1}
$$

$$
= O \left( \frac{|I_{j_0}|}{(d_n |I_{j_0}|)^2} \right) + \sum_{1,1}
$$

$$
= O \left( \frac{n\sqrt{1-u}}{d_n^2} \right) + \sum_{1,1}, \tag{10.13}
$$

where

$$\sum_{1,1} = \sum_{\substack{j=j_0-L, j\neq j_0}}^{j_0+L} \frac{u-\xi_j}{\tau_j} = \sum_{k=1}^{L} \left\{ \frac{u-\xi_{j_0+k}}{\tau_{j_0+k}} + \frac{u-\xi_{j_0-k}}{\tau_{j_0-k}} \right\}. \qquad (10.14)$$

Recall here $\tau_j$ is defined by (10.5). Note that by (10.6) and (10.8),

$$\tau_{j_0 \pm k} \sim \left( k^2 + d_n^2 \right) \left| I_{j_0} \right|^2.$$

Here

$$\frac{u-\xi_{j_0+k}}{\tau_{j_0+k}} + \frac{u-\xi_{j_0-k}}{\tau_{j_0-k}}$$

$$= \frac{1}{\tau_{j_0+k}\tau_{j_0-k}} \left[ \begin{array}{l} \left(u-\xi_{j_0+k}\right)\left[\left(u-\xi_{j_0-k}\right)^2 + \left(d_n\left|I_{j_0-k}\right|\right)^2\right] \\ + \left(u-\xi_{j_0-k}\right)\left[\left(u-\xi_{j_0+k}\right)^2 + \left(d_n\left|I_{j_0+k}\right|\right)^2\right] \end{array} \right]$$

$$= \frac{1}{\tau_{j_0+k}\tau_{j_0-k}} \left[ \begin{array}{l} \left(u-\xi_{j_0+k}\right)\left(u-\xi_{j_0-k}\right)\left(\left(u-\xi_{j_0-k}\right) + \left(u-\xi_{j_0+k}\right)\right) \\ + \left(d_n\left|I_{j_0-k}\right|\right)^2\left(\left(u-\xi_{j_0+k}\right) + \left(u-\xi_{j_0-k}\right)\right) \\ + d_n^2\left(u-\xi_{j_0-k}\right)\left(\left|I_{j_0+k}\right|^2 - \left|I_{j_0-k}\right|^2\right) \end{array} \right]$$

$$= \frac{1}{\tau_{j_0+k}\tau_{j_0-k}} \left[ O\left(k^2\left|I_{j_0}\right|^2 \rho_k\right) + O\left(d_n^2\left|I_{j_0}\right|^2 \rho_k\right) \right.$$

$$\left. + O\left(d_n^2\left|I_{j_0}\right|^2 k\left|\left|I_{j_0+k}\right| - \left|I_{j_0-k}\right|\right|\right) \right]$$

(by (10.9) and (10.8))

$$\leq \frac{C}{\left|I_{j_0}\right|^4\left(k^2+d_n^2\right)^2}\left[\left(k^2+d_n^2\right)\left|I_{j_0}\right|^2 + d_n^2\left|I_{j_0}\right|^2\right]\rho_k$$

$$\leq \frac{C\rho_k}{\left|I_{j_0}\right|^2\left(k^2+d_n^2\right)},$$

by (10.10). Substituting in (10.14), and using the definition (10.9) of $\rho_k$,

$$\sum_{1,1} \leq \frac{C}{|I_{j_0}|} \sum_{k=1}^{L} \frac{1}{k^2 + d_n^2} \left\{ 1 + \frac{k^2}{n(1-u)^{3/2}} + k\omega \left( \frac{k}{n\sqrt{1-u}} \right) \right\}$$

$$\leq Cn\sqrt{1-u} \int_0^L \frac{dx}{x^2 + d_n^2} + \frac{C}{1-u} \int_0^L \frac{x^2}{x^2 + d_n^2} dx$$

$$+ Cn\sqrt{1-u}\,\omega \left( \frac{L}{n\sqrt{1-u}} \right) \int_0^L \frac{x}{x^2 + d_n^2} dx$$

$$\leq C\frac{n}{d_n} \int_0^\infty \frac{ds}{s^2+1} + \frac{C}{1-u} d_n \int_0^{L/d_n} \frac{s^2}{s^2+1} ds$$

$$+ Cn\sqrt{1-u}\,\omega \left( \frac{L}{n\sqrt{1-u}} \right) \int_0^{L/d_n} \frac{s}{s^2+1} ds$$

$$\leq C\frac{n}{d_n} + C\frac{L}{1-u} + Cn\sqrt{1-u}\,\omega \left( \frac{L}{n\sqrt{1-u}} \right) \log d_n$$

$$\leq C\frac{n}{d_n} + C\frac{n}{d_n^4} + C\frac{n}{d_n^2} \log d_n \leq C\frac{n}{d_n},$$

by (9.4) and (9.5) and since $L = d_n^2$. Together with (10.13), this gives the result.   □
Finally, we handle the central integral:

**Lemma 10.5.**

$$|J_1| = \left| nPV \int_{t_{j_0}-L}^{t_{j_0}+L+1} \frac{\sigma(t)}{u-t} dt \right| \leq C\frac{n}{d_n}. \tag{10.15}$$

*Proof.* Now

$$J_1 = n \int_{t_{j_0}-L}^{t_{j_0}+L+1} \frac{\sigma(t) - \sigma(u)}{u-t} dt - n\sigma(u) PV \int_{t_{j_0}-L}^{t_{j_0}+L+1} \frac{dt}{t-u}$$

$$= J_{11} - J_{12}.$$

Here, using (9.2), (10.6), and (10.8),

$$|J_{11}| = \left| n \int_{t_{j_0}-L}^{t_{j_0}+L+1} \frac{\sqrt{1-t^2} - \sqrt{1-u^2}}{u-t} h(t)\, dt \right.$$

$$\left. + n\sqrt{1-u^2} \int_{t_{j_0}-L}^{t_{j_0}+L+1} \frac{h(t) - h(u)}{t-u}\, dt \right|$$

$$\le Cn \int_{t_{j_0}-L}^{t_{j_0}+L+1} |u-t|^{-1/2}\, dt + Cn \int_{t_{j_0}-L}^{t_{j_0}+L+1} \frac{\omega(|t-u|)}{|t-u|}\, dt$$

$$\le Cn \left(L\,|I_{j_0}|\right)^{1/2} + Cn \int_0^{CL|I_{j_0}|} \frac{\omega(s)}{s}\, ds$$

$$\le Cn \left(\frac{d_n^2}{n\sqrt{1-u}}\right)^{1/2} + Cn \int_0^{Cd_n^2/(n\sqrt{1-u})} \frac{\omega(s)}{s}\, ds \le C\frac{n}{d_n}, \tag{10.16}$$

by (10.1) and (9.4). Next,

$$J_{12} = n\sigma(u) \log \left| \frac{t_{j_0+L+1} - u}{u - t_{j_0-L}} \right|$$

$$= n\sigma(u) \log \left| 1 + \frac{(t_{j_0+L+1} - u) + (t_{j_0-L} - u)}{u - t_{j_0-L}} \right|$$

$$= n\sigma(u) \log \left( 1 + O\left( \frac{\left| (u - \xi_{j_0-L}) + (u - \xi_{j_0+L}) \right| + |I_{j_0}|}{L\,|I_{j_0}|} \right) \right)$$

$$= O\left( n\sigma(u) \frac{\rho_L + |I_{j_0}|}{L\,|I_{j_0}|} \right) \quad \text{(recall (10.9))}$$

$$= O\left( \frac{n\sigma(u)}{L} \left\{ 1 + \frac{L^2}{n(1-u)^{3/2}} \right\} + n\sigma(u)\,\omega\left( \frac{L}{n\sqrt{1-u}} \right) \right)$$

$$= O\left( \frac{n}{L} \right) + O\left( \frac{L}{1-u} \right) + n\sigma(u)\,\omega\left( \frac{d_n^2}{n\sqrt{1-u}} \right) = O\left( \frac{n}{d_n^2} \right),$$

by (9.4) and (9.5). Then together with (10.16), this gives (10.15).                    □

*Proof of Theorem* 10.1.  This follows by combining Lemmas 10.3, 10.4, and 10.5. Indeed, from (10.3),

$$\frac{1}{n} \left| \frac{d}{du} \log \left\{ |R_n(u)| \, e^{n V^{\sigma_n}(u)} \right\} \right|$$

$$= \frac{1}{n} \left| \Gamma_n'(u) \right|$$

$$= \frac{1}{n} \left| \sum_{j : |j - j_0| \le L} \Gamma_{n,j}^{(1)}(u) + \sum_{j : |j - j_0| > L} \Gamma_{n,j}^{(1)}(u) \right|$$

$$\le \frac{1}{n} \left( |\Sigma_1| + |J_1| + \sum_{j : |j - j_0| > L} \left| \Gamma_{n,j}^{(1)}(u) \right| \right)$$

$$\le \frac{C}{d_n} + \frac{C}{d_n} + \frac{C}{d_n^2} \le \frac{C}{d_n},$$

by Lemmas 10.3, 10.4, and 10.5.                                                           □

# Chapter 11
# Weighted Polynomial Approximations

In this chapter, we turn the weighted polynomial approximations of the previous chapter into ones suitable for establishing asymptotics. Recall that a Bernstein-Szegő weight has the form $w_{n,B}^2$, where

$$w_{n,B}(t) = \frac{\left(1 - t^2\right)^{1/4}}{\sqrt{S_n(t)}}.$$  (11.1)

Here $S_n$ is a polynomial of degree $\leq 2n$, that is positive in $[-1, 1]$, except possibly for simple zeros at $\pm 1$. We first state our three main results and then turn to the proofs. Throughout we assume that $\{Q_n\} \in \mathcal{Q}$.

**Theorem 11.1.** *Let* $\tau_1 \in (0, \frac{\alpha}{50}]$. *There exist for n large enough, polynomials* $S_n$ *of degree* $\leq 2n - n^{1/3-2\tau_1}$, *positive in* $[-1, 1]$, *such that if* $w_{n,B}$ *is given by* (11.1), *and*

$$\psi_n(t) = e^{-nQ_n(t)}/w_{n,B}(t),$$  (11.2)

*then*
*(a)*

$$|\psi_n(t) - 1| \leq Cn^{-\tau_1}, \quad |t| \leq 1 - 4n^{-\frac{2}{3}+12\tau_1}.$$  (11.3)

*(b)*

$$n^{-2\tau_1}\left(1 - t^2\right)^{1/4} \leq \psi_n(t) \leq 1, \quad t \in [-1, 1].$$  (11.4)

*(c) For* $|t| \leq 1 - 4n^{-\frac{2}{3}+12\tau_1}$,

$$|\psi_n'(t)| \leq Cn^{1-\tau_1}.$$  (11.5)

© The Author(s) 2018
E. Levin, D.S. Lubinsky, *Bounds and Asymptotics for Orthogonal Polynomials for Varying Weights*, SpringerBriefs in Mathematics,
https://doi.org/10.1007/978-3-319-72947-3_11

*(d) For* $t \in (-1, 1)$

$$\left| \frac{\psi_n'(t)}{\psi_n(t)} \right| \le C \left( \frac{n^{1+4\tau_1}}{(1-t^2)^{1/2}} + \frac{n^{2/3+4\tau_1}}{1-t^2} \right). \tag{11.6}$$

*(e)*

$$\int_{-1}^{1} \frac{|\log \psi_n(t)|}{\sqrt{1-t^2}} dt \le C n^{-\tau_1}. \tag{11.7}$$

*(f) For* $|x| \le 1 - n^{-\tau_1}$,

$$\int_{-1}^{1} \frac{\left| 1 - \left( \frac{\psi_n(t)}{\psi_n(x)} \right)^2 \right| \psi_n(t)^{-1}}{|x - t|} \frac{dt}{\sqrt{1-t^2}} \le C n^{-\tau_1/3}. \tag{11.8}$$

*(g) For* $|x| \le 1 - n^{-\tau_1}$,

$$\int_{-1}^{1} \left| \frac{\log \psi_n^{-2}(t) - \log \psi_n^{-2}(x)}{t - x} \right| \frac{dt}{\sqrt{1-t^2}} \le C n^{-\tau_1/3}. \tag{11.9}$$

*(h) For* $1 - n^{-\tau_1} \le |x| < 1$,

$$\int_{-1}^{1} \frac{\left| 1 - \left( \frac{\psi_n(t)}{\psi_n(x)} \right)^2 \right| \psi_n(t)^{-1}}{|x - t|} \frac{dt}{\sqrt{1-t^2}} \le C n^{5\tau_1} (1 - |x|)^{-5/4}. \tag{11.10}$$

*(i) For* $|x| \le 1 - n^{-\tau_1}$,

$$\int_{-1}^{1} \left| \frac{\frac{\psi_n'(t)}{\psi_n(t)}(1-t^2) - \frac{\psi_n'(x)}{\psi_n(x)}(1-x^2)}{t - x} \right| \frac{dt}{\sqrt{1-t^2}} \le C n^{1-\tau_1/3}. \tag{11.11}$$

The next two results involve one-sided approximation:

**Theorem 11.2.** *Let* $\tau_1 \in (0, \frac{\alpha}{50}]$. *Let* $A > 0$ *and*

$$\rho_n = 1 + \left( A \frac{\log n}{n} \right)^{2/3}, \quad n \ge 2. \tag{11.12}$$

*There exist for n large enough, polynomials* $S_n$ *of degree* $\le 2n - n^{1/3-2\tau_1}$, *positive in* $[-1, 1]$, *such that if* $w_{n,B}$ *is given by* (11.1), *and*

$$\hat{\psi}_n(t) = e^{-nQ_n(\rho_n t)} / w_{n,B}(t), \tag{11.13}$$

*then*
*(a)*

$$\left| \hat{\psi}_n(t) - 1 \right| \le Cn^{-\tau_1}, \quad |t| \le 1 - n^{-\frac{2}{3} + 12\tau_1}. \tag{11.14}$$

*(b)*

$$\hat{\psi}_n(t) \le 1, \quad t \in [-1, 1]. \tag{11.15}$$

*(c)*

$$\int_{-1}^{1} \frac{\left| \log \hat{\psi}_n(t) \right|}{\sqrt{1 - t^2}} dt \le Cn^{-\tau_1}. \tag{11.16}$$

**Theorem 11.3.** *There exist for n large enough, polynomials $S_n$ of degree $\le 2n - n^{1/3 - 2\tau_1}$, positive in $[-1, 1]$, such that if $w_{n,B}$ and $\psi_n$ are given by (11.1) and (11.2) respectively, then*

$$\psi_n(t) \ge 1, \quad t \in [-1, 1], \tag{11.17}$$

*and*

$$\int_{-1}^{1} \frac{\left| \log \psi_n(t) \right|}{\sqrt{1 - t^2}} dt \le Cn^{-\tau_1/3}. \tag{11.18}$$

The procedure to prove these theorems is somewhat technical. We use the polynomials $\{R_n\}$ from Theorem 9.1. Unfortunately these are often too large or too small near $\pm 1$, so we have to multiply by damping polynomials, or add weighted polynomials that are bounded below. This increases the degree of the approximating polynomials, and we then have to scale $Q_n$ to adjust the degree.

We start by reformulating parts of Theorems 9.1 and 10.1:

**Lemma 11.4.** *There exists $n_0 > 0$ with the following property: for $n \ge n_0$, and $\tau_1 \in (0, \frac{\alpha}{50}]$, there exist polynomials $R_n^\#$ with no real zeros, of degree $\le n$, such that*
*(a)*

$$\left| R_n^\#(x) \right| e^{-nQ_n(x)} = 1 + O(n^{-\tau_1}), \quad |x| \le 1 - n^{-\frac{2}{3} + 8\tau_1}. \tag{11.19}$$

*(b)*

$$\exp\left( C_1 n^{\tau_1} \log n \right) \ge \left| R_n^\#(x) \right| e^{-nQ_n(x) - nU_n(x)} \ge \exp\left( -\pi n^{\tau_1} \right), \quad x \in I_n. \tag{11.20}$$

*(c)*

$$\exp\left( C_1 n^{\tau_1} \log n \right) \ge \left| R_n^\#(x) \right| e^{-nQ_n(x)} \ge \exp\left( -\pi n^{\tau_1} \right), \quad x \in [-1, 1]. \tag{11.21}$$

*(d)*

$$\left| \frac{d}{dx} \left\{ \log \left( |R_n^{\#}(x)| \, e^{-nQ_n(x)} \right) \right\} \right| \leq C n^{1-\tau_1}, \quad |x| \leq 1 - n^{-\frac{2}{3}+8\tau_1}. \tag{11.22}$$

*Proof.* We repeatedly use the fact that $\tau_1 \in \left( 0, \frac{\alpha}{50} \right)$. Set $d_n = [n^{\tau_1}]$ and $\sigma = \hat{\sigma}_{Q_n}$ in Theorem 9.1. We apply that theorem and Theorem 10.1 with $\omega(s) = s^{\alpha/2}$, which is permissible in view of (3.14). Let

$$R_n^{\#}(x) = R_n(x) \, e^{nFQ_n} e^{-d_n\pi}.$$

Observe that from (3.11),

$$\log \left( |R_n^{\#}(x)| \, e^{-nQ_n(x)} \right)$$

$$= \log \left( |R_n(x)| \, e^{nV^{\sigma Q_n}(x)} \right) + nU_n(x) - d_n\pi. \tag{11.23}$$

(a) If $|x| \leq 1 - n^{-\frac{2}{3}+8\tau_1}$, then

$$n(1 - |x|)^{3/2} \geq n^{12\tau_1} \geq d_n^6$$

for large enough $n$, and

$$\left( \frac{n(1 - |x|)^{1/2}}{d_n^4} \right)^{\alpha/2} \geq n^{\alpha/3} \geq n^{2\tau_1} \geq d_n^2.$$

So both (9.4) and (9.5) hold. Then as $U_n = 0$ in $(-1, 1)$, (9.6) of Theorem 9.1 and (11.23) give

$$\log \left( |R_n^{\#}(x)| \, e^{-nQ_n(x)} \right) = O\left( d_n^{-1} \right) = O\left( n^{-\tau_1} \right),$$

so we have (11.19).

(b) This follows directly from (9.3) and (9.7) of Theorem 9.1 and (11.23).
(c) This follows from (b) as $U_n = 0$ in $[-1, 1]$.
(d) Observe first that (10.1) holds with $\omega(s) = s^{\alpha/2}$ and $u$ replaced by $x$. Indeed,

$$\int_0^{d_n^2/(n\sqrt{1-|x|})} \frac{\omega(s)}{s} ds \leq C \left( \frac{d_n^2}{n\sqrt{1-|x|}} \right)^{\alpha/2} \leq C d_n^{-2},$$

as above. By (11.23), for $|x| \leq 1 - n^{-\frac{2}{3}+8\tau_1}$, and then Theorem 10.1,

$$\frac{d}{dx} \left\{ \log \left( |R_n^{\#}(x)| \, e^{-nQ_n(x)} \right) \right\} = \frac{d}{dx} \left\{ \log \left( |R_n(x)| \, e^{nV^{\sigma Q_n}(x)} \right) \right\} = O\left( n^{1-\tau_1} \right).$$

$$\square$$

Next, we use damping polynomials to get around the fact that our discretized polynomials may be too large near $\pm 1$.

**Lemma 11.5.** *There exists $n_0$ and for $n \geq n_0$, and $\tau_1 \in (0, \frac{\alpha}{50}]$, polynomials $S_n$ with no zeros in $[-1, 1]$, of degree $\leq n + n^{1/3-2\tau_1}$ such that*
*(a)*

$$e^{-2n^{2\tau_1}+nU_n(x)} \leq |S_n(x)| e^{-nQ_n(x)} \leq 1, \quad x \in I_n. \tag{11.24}$$

*(b)*

$$|S_n(x)| e^{-nQ_n(x)} = 1 + O(n^{-\tau_1}), \quad |x| \leq 1 - 3n^{-\frac{2}{3}+8\tau_1}. \tag{11.25}$$

*(c)*

$$\left| \frac{d}{dx} \log \left\{ \left| S_n(x) e^{-nQ_n(x)} \right| \right\} \right| \leq Cn^{1-\tau_1}, \quad |x| \leq 1 - 3n^{-\frac{2}{3}+8\tau_1}. \tag{11.26}$$

*Proof.* Let

$$m = m(n) = \left[ \frac{1}{2} n^{1/3-2\tau_1} \right] \text{ and } \zeta_m = 2n^{-\frac{2}{3}+8\tau_1}. \tag{11.27}$$

Note that then $m^2 \zeta_m \sim n^{4\tau_1} \to \infty$ as $n \to \infty$. By Theorem 7.5 in [25, p. 172], there exist polynomials $P_m$ of degree $\leq m$ such that

$$0 < P_m(x) \leq \exp\left(-Cm\sqrt{\zeta_m}\right)$$

$$\leq \exp\left(-Cn^{2\tau_1}\right), \quad 1 - n^{-\frac{2}{3}+8\tau_1} \leq |x| \leq 1; \tag{11.28}$$

$$|1 - P_m(x)| \leq \exp\left(-Cm\sqrt{\zeta_m}\right)$$

$$\leq \exp\left(-Cn^{2\tau_1}\right), \quad |x| \leq 1 - 3n^{-\frac{2}{3}+8\tau_1}. \tag{11.29}$$

$$0 < P_m(x) \leq 1, \quad x \in [-1, 1]. \tag{11.30}$$

Then by Bernstein's inequality, for $|x| \leq 1 - 3n^{-\frac{2}{3}+8\tau_1}$

$$|P'_m(x)| \leq \frac{m}{\sqrt{1-x^2}} \leq Cn^{2/3-6\tau_1}. \tag{11.31}$$

Let $R_n^{\#}$ be as in the last lemma. Define

$$\breve{S}_n(x) = R_n^{\#}(x) \left[ P_m^2(x) + e^{-n^{2\tau_1}} \right],$$

a polynomial of degree $\leq n + n^{1/3-2\tau_1}$.

(a) Firstly for $|x| \leq 1 - 3n^{-\frac{2}{3}+8\tau_1}$, (11.19) and (11.29) give

$$\log \left( \left| \hat{S}_n (x) \right| e^{-nQ_n(x)} \right)$$

$$= \log \left( \left| R_n^\# (x) \right| e^{-nQ_n(x)} \right) + \log \left[ P_m^2 (x) + e^{-n^2\tau_1} \right]$$

$$= O(n^{-\tau_1}) + O\left( \exp\left(-Cn^{2\tau_1}\right) \right) = O(n^{-\tau_1}). \tag{11.32}$$

Next, for $1 - n^{-\frac{2}{3}+8\tau_1} \leq |x| \leq 1$, (11.21) and (11.28) give

$$\log \left( \left| \hat{S}_n (x) \right| e^{-nQ_n(x)} \right) \leq C_0 n^{\tau_1} \log n - Cn^{2\tau_1} \leq -C_1 n^{2\tau_1}.$$

In the remaining range $1 - 3n^{-\frac{2}{3}+8\tau_1} \leq |x| \leq 1 - n^{-\frac{2}{3}+8\tau_1}$, we obtain from (11.19) and (11.30),

$$\log \left( \left| \hat{S}_n (x) \right| e^{-nQ_n(x)} \right) \leq O(n^{-\tau_1}).$$

Combining these last three inequalities, we obtain for $x \in [-1, 1]$,

$$\left| \hat{S}_n (x) \right| e^{-nQ_n(x)} \leq 1 + Cn^{-\tau_1}.$$

Next, we apply Theorem 4.2(a) with $T = 1$ and $S = n^{1/3-2\tau_1}$. That result applied to $\hat{S}_n$, which has degree $\leq nT + S$,

$$\left\| \hat{S}_n e^{-nQ_n} \right\|_{L_\infty (I_n \setminus [-1,1])} \leq e^{C(S+1)n^{-1/2}} \left\| \hat{S}_n e^{-nQ_n} \right\|_{L_\infty[-1,1]}$$

$$\leq e^{Cn^{-1/6-2\tau_1}} (1 + Cn^{-\tau_1}) \leq 1 + C_1 n^{-\tau_1}. \tag{11.33}$$

Defining

$$S_n = \hat{S}_n / (1 + C_1 n^{-\tau_1}),$$

we obtain the right inequality in (11.24). For the left, we use the lower bound in (11.20): for $x \in I_n$,

$$\left| \hat{S}_n (x) \right| \geq \left| R_n^\# (x) \right| e^{-nQ_n(x)} e^{-n^2\tau_1} \geq e^{-n^2\tau_1 - \pi n^{\tau_1} + nU_n(x)}.$$

Then the lower bound (11.24) follows for large enough $n$.

(b) This follows directly from (11.32).

(c) For $|x| \leq 1 - 3n^{-\frac{2}{3}+8\tau_1}$,

$$\left| \frac{d}{dx} \log \left\{ \left| S_n(x) e^{-nQ_n(x)} \right| \right\} \right|$$

$$\leq \left| \frac{d}{dx} \log \left\{ \left| R_n^{\#}(x) e^{-nQ_n(x)} \right| \right\} \right| + \left| \frac{d}{dx} \log \left\{ P_m(x)^2 + e^{-n^{2\tau_1}} \right\} \right|$$

$$\leq C n^{1-\tau_1} + C n^{2/3-6\tau_1},$$

by (11.22) and (11.31). Then (11.26) follows.                                □

The degree of $S_n$ in the last lemma exceeds $n$. We now manipulate the definition of $Q_n$ to reduce its degree.

**Lemma 11.6.** *Let* $\Delta \in \left(0, \frac{2}{3}\right)$. *There exists for* $n \geq n_0$, *and* $\tau_1 \in \left(0, \frac{\alpha}{50}\right]$, *polynomials* $X_n$ *of degree* $\leq n - n^{1/3-2\tau_1}$ *such that*
*(a)*

$$|X_n(x)| e^{-nQ_n(x)} \leq 1, \quad x \in I_n. \tag{11.34}$$

*(b)*

$$|X_n(x)| e^{-nQ_n(x)} \geq e^{-Cn^{3\tau_1}-Cn^{3\Delta/2}}, \quad |x| \leq 1 + n^{-2/3+\Delta}. \tag{11.35}$$

*(c)*

$$|X_n(x)| e^{-nQ_n(x)} = 1 + O\left(n^{-\tau_1}\right), \quad |x| \leq 1 - 4n^{-\frac{2}{3}+8\tau_1}. \tag{11.36}$$

*(d)*

$$\left| \frac{d}{dx} \left\{ \log \left( |X_n(x)| e^{-nQ_n(x)} \right) \right\} \right| \leq C n^{1-\tau_1}, \quad |x| \leq 1 - 4n^{-\frac{2}{3}+8\tau_1}. \tag{11.37}$$

*Proof.* For large enough $n$, let

$$m = m(n) = n - 2\left[n^{\frac{1}{3}-2\tau_1}\right] - 1 \text{ and } r_n = \frac{m}{n}. \tag{11.38}$$

Note that

$$m + m^{1/3-2\tau_1} \leq n - n^{1/3-2\tau_1}$$

and

$$1 - r_n \sim n^{-\frac{2}{3}-2\tau_1}. \tag{11.39}$$

Set

$$Q_m^{\#}(t) = r_n^{-1} Q_n\left(L_{n,r_n}^{[-1]}(t)\right), \quad t \in L_{n,r_n}(I_n) = I_m^{\#}.$$

Since $L_{n,r_n}^{[-1]}$ maps $[-1, 1]$ onto $[a_{-n,r_n}, a_{n,r_n}]$, and since $Q_m^{\#}$ is convex, this external field has equilibrium measure with support $[-1, 1]$. See the proof of the general case of Theorem 7.1, especially (7.19) and (7.20), where this is established in detail. Since $r_n \in \left[r_0, \frac{1}{r_0}\right]$ for large enough $n$, we can apply the conclusions of Lemmas 3.3 and 3.6. It follows that we can apply the conclusions of Lemma 11.5 to the fields $\{Q_m^{\#}\}$ on $\{I_m^{\#}\}$. It is conceivable that not every positive integer is included in the sequence $\{m(n)\}$, or that finitely many are repeated, but this is not an issue. Moreover, $\{Q_m^{\#}\}$ satisfy all the conditions of Definition 1.1, as in the proof of the general case of Theorem 7.1. Thus there exist polynomials $S_m^{\#}$ of degree $\le m + m^{1/3-2\tau_1} \le n - n^{1/3-2\tau_1}$ with the properties that

$$e^{-2m^{2\tau_1}+mU_m^{\#}(t)} \le \left|S_m^{\#}(t)\right| e^{-mQ_m^{\#}(t)} \le 1, \quad t \in I_m^{\#}; \tag{11.40}$$

$$\left|S_m^{\#}(t)\right| e^{-mQ_m^{\#}(t)} = 1 + O\left(m^{-\tau_1}\right), \quad |t| \le 1 - 3m^{-\frac{2}{3}+8\tau_1}; \tag{11.41}$$

and

$$\left|\frac{d}{dt}\log\left\{\left|S_m^{\#}(t)\right| e^{-mQ_m^{\#}(t)}\right\}\right| \le Cm^{1-\tau_1}, \quad |t| \le 1 - 3m^{-\frac{2}{3}+8\tau_1}. \tag{11.42}$$

Here $U_m^{\#}(t) = -\left(V^{\sigma_{Q_m^{\#}}}(t) + Q_m^{\#}(t) - F_{Q_m}^{\#}\right)$ is the $Q_m^{\#}$ analogue of $U_n$ for $Q_n$. Now let

$$X_n(x) = S_m^{\#}\left(L_{n,r_n}(x)\right),$$

a polynomial of degree $\le n - n^{1/3-2\tau_1}$. Note that for $t = L_{n,r_n}(x)$,

$$mQ_m^{\#}\left(L_{n,r_n}(x)\right) = nQ_n(x),$$

and

$$|X_n(x)| e^{-nQ_n(x)} = \left|S_m^{\#}(t)\right| e^{-mQ_m^{\#}(t)}.$$

(a) Now (11.40) gives

$$|X_n(x)| e^{-nQ_n(x)} \le 1, \quad x \in I_n,$$

so we have (11.34).

(b) By (11.40), and the definition of $X_n$, for $x \in I_n$,

$$|X_n(x)| e^{-nQ_n(x)} \ge e^{-2m^{2\tau_1}+mU_m^{\#}\left(L_{n,r_n}(x)\right)}.$$

For $x \in L_{n,r_n}^{[-1]}([-1, 1])$, we then immediately obtain (11.35). Now consider $L_{n,r_n}^{[-1]}(1) \le x \le 1 + n^{-2/3+\Delta}$. As $Q_m^{\#}$ has equilibrium density with support

$[-1, 1]$, (3.34) of Lemma 3.6 gives

$$mU_m^{\#}(L_{n,r_n}(x)) \geq -Cn(L_{n,r_n}(x) - 1)^{3/2}$$

$$= -Cn\left(\frac{x - a_{n,r_n}}{\delta_{n,r_n}}\right)^{3/2}$$

$$\geq -Cn((x-1) + (1 - r_n))^{3/2}$$

$$\geq -Cn^{3\Delta/2} - Cn^{-3\tau_1},$$

by (3.26) and (11.39) and since $\delta_{n,r_n} \sim 1$. Then (11.35) follows.
(c) Setting $t = L_{n,r_n}(x)$ in (11.41) gives

$$|X_n(x)|e^{-nQ_n(x)} = 1 + O(n^{-\tau_1}),$$

provided

$$|L_{n,r_n}(x)| \leq 1 - 3m^{-\frac{2}{3}+8\tau_1} \Leftrightarrow x \in L_{n,r_n}^{[-1]}\left[-1 + 3m^{-\frac{2}{3}+8\tau_1}, 1 - 3m^{-\frac{2}{3}+8\tau_1}\right].$$

Here

$$1 - L_{n,r_n}^{[-1]}\left(1 - 3m^{-\frac{2}{3}+8\tau_1}\right) = 1 - a_{n,r_n} + 3m^{-\frac{2}{3}+8\tau_1}\delta_{n,r_n}$$

$$= O\left(n^{-\frac{2}{3}-2\tau_1}\right) + 3n^{-\frac{2}{3}+8\tau_1}\left(1 + O\left(n^{-\frac{2}{3}-2\tau_1}\right)\right),$$

by (11.39) and (3.27). Thus if $|x| \leq 1 - 4n^{-2/3+8\tau_1}$, this is satisfied for large enough $n$, and we obtain (11.36).
(d) Finally, $\frac{dt}{dx} = \frac{d}{dx}L_{n,r_n}(x) = \frac{1}{\delta_{n,r_n}} \sim 1$, so for the range in (11.37), we obtain from (11.42),

$$\frac{d}{dx}\left\{\log\left(|X_n(x)|e^{-nQ_n(x)}\right)\right\} = \frac{d}{dt}\left\{\log\left(|S_m^{\#}(t)|e^{-mQ_m^{\#}(t)}\right)\right\}\frac{1}{\delta_{n,r_n}}$$

$$= O\left(n^{1-\tau_1}\right).$$

$\square$

Next, we use Christoffel functions to construct weighted polynomials that are bounded above and below:

**Lemma 11.7.** *Let $A > 0$. There exist $C_1, C_2 > 0$ and polynomials $E_n$ of degree $\leq 2n - 3n^{1/3}$ such that for $n \geq 1$ and $|x| \leq 1 + An^{-2/3}$,*

$$C_1 \leq E_n(x)e^{-2nQ_n(x)} \leq C_2. \tag{11.43}$$

*Proof.* Recall from Theorem 5.1 that if $N = N(n) = n - [2n^{1/3}]$, then for $|x| \le 1 + An^{-2/3}$ and $n \ge 1$,

$$\lambda_N^{-1}\left(e^{-2nQ_n}, x\right) e^{-2nQ_n(x)} \sim n\sqrt{1 - |x| + n^{-2/3}}.$$

Now consider the Legendre weight

$$v(x) = 1, \quad x \in [-1, 1].$$

Its Christoffel function $\lambda_m(v, x)$ satisfies uniformly in $m$ and $|x| \le 1 + Am^{-2}$, [39, p. 108, Lemma 5]

$$\lambda_m^{-1}(v, x) \sim m\left(1 - |x| + m^{-2}\right)^{-1/2}.$$

Then we choose

$$E_n(x) = \left(\frac{1}{n}\lambda_N^{-1}\left(e^{-2nQ_n}, x\right)\right)\left(\frac{1}{n^{1/3}}\lambda_{[\frac{1}{2}n^{1/3}]}^{-1}(v, x)\right).$$

It is easily seen from the above relations that $E_n$ does the job.    □

Our last lemma involves one-sided approximations to $(1 - x^2)^{1/2}$. Surprisingly it involves far more work than one would expect.

**Lemma 11.8.**   *(I) Let $\varepsilon \in (0, \frac{1}{2})$. There exist for large enough $m$, polynomials $Y_m$ of degree $\le 4m$ such that*

$$\frac{1}{2}m^{-\varepsilon/4}\left(1 - t^2\right) \le Y_m(t) \le \left(1 - t^2\right)^{1/2}, \quad |t| \le 1; \tag{11.44}$$

$$Y_m(t) = \left(1 - t^2\right)^{1/2}\left(1 + O\left(m^{-\varepsilon/4}\right)\right), \quad |t| \le 1 - m^{-2+\varepsilon}; \tag{11.45}$$

*and*

$$\left|Y_m'(t)\right| \le Cm^{2-\varepsilon/2}, \quad |t| \le 1 - m^{-2+\varepsilon}. \tag{11.46}$$

*(II) Similarly, there exist polynomials $Y_m$ of degree $\le 2m$ satisfying (11.45), (11.46), and for $|t| < 1$,*

$$(1 - t^2)^{1/2} + \frac{C}{m^3(1 - t^2)} \ge Y_m(t) \ge \left(1 - t^2\right)^{1/2}. \tag{11.47}$$

*Proof.* (I) Let $S_{2m}$ be the polynomial of degree $\le 2m - 1$ interpolating the function $f(t) = \left(1 - t^2\right)^{-3/4}$ at the $2m$ zeros of $T_m^2$ in $(-1, 1)$. Here $T_m$ is the usual Chebyshev polynomial of degree $m$. Since $T_m^2$ is even, as is $f$, the degree of $S_{2m}$ is actually

$\leq 2m-2$. By the error formula for Lagrange interpolation [33, p. 54], for $t \in (-1, 1)$, there exists $\xi \in (-1, 1)$ such that

$$f(t) - S_{2m}(t) = \frac{f^{(2m)}(\xi)}{(2m)!} T_m^2(t).$$  (11.48)

Since $f^{(2m)} \geq 0$, we have

$$S_{2m}(t) \leq (1 - t^2)^{-3/4}, \quad t \in (-1, 1).$$

Let

$$R_m(t) = (1 - t^2)^2 S_{2m}^2(t),$$

a polynomial of degree $\leq 4m$. Then we have

$$0 \leq R_m(t) \leq (1 - t^2)^{1/2}, \quad t \in (-1, 1).$$  (11.49)

Next, we use the error formula for Lagrange interpolation: if $\Gamma$ is a simple closed positively oriented contour enclosing $(-1, 1)$ and passing through $\pm 1$, then by taking limits in the usual error formula for Lagrange interpolation to deal with the integrable singularities at $\pm 1$, [33, p. 55],

$$f(x) - S_{2m}(x) = \frac{1}{2\pi i} \int_\Gamma \frac{f(t)}{t - x} \left(\frac{T_m(x)}{T_m(t)}\right)^2 dt, \quad x \in (-1, 1).$$

Here the branch of $f(t) = (1 - t^2)^{-3/4}$ is taken so that $f$ is real valued in $(-1, 1)$ and analytic in $\mathbb{C} \backslash ((-\infty, -1] \cup [1, \infty))$. Since the integrand is analytic as a function of $t$ in $\mathbb{C} \backslash ((-\infty, -1] \cup [1, \infty))$ and outside $\Gamma$, and is $O(t^{-2m-7/4})$ at $\infty$, we can deform the contour $\Gamma$ into two "loops" $\Gamma_+$ and $\Gamma_-$ around $\pm 1$. Here the "loop" $\Gamma_+$ consists of a line from $\infty$ to 1 in the lower half-plane, encircles 1 clockwise, and then is a line from 1 to $\infty$ in the upper-half plane. $\Gamma_-$ is described similarly. Then we see that for $x \in (-1, 1)$,

$$f(x) - S_{2m}(x) = \frac{1}{2\pi i} \left(\int_1^\infty + \int_{-\infty}^{-1}\right) \frac{f(s)_+ - f(s)_-}{s - x} \left(\frac{T_m(x)}{T_m(s)}\right)^2 ds.$$

Here $f(s)_\pm$ are the boundary values of $f$ from the upper and lower half planes. Using evenness, we deduce that for $x \in [0, 1)$,

$$|f(x) - S_{2m}(x)| \leq \frac{2}{\pi} \int_1^\infty \frac{(s^2 - 1)^{-3/4}}{s - x} \frac{ds}{T_m(s)^2}.$$  (11.50)

Next, we use the bound

$$T_m(s) \geq \frac{s^m}{2} e^{\frac{m}{2}(1-s^{-1})^{1/2}}, \quad s \in [1,\infty).$$

Indeed, for $s \in [1,\infty)$,

$$T_m(s) = \frac{1}{2}\left(\left(s + \sqrt{s^2-1}\right)^m + \left(s - \sqrt{s^2-1}\right)^m\right)$$

$$\geq \frac{s^m}{2}\left(1 + \sqrt{1-s^{-2}}\right)^m$$

$$= \frac{s^m}{2}\exp\left(m\log\left(1 + \sqrt{1-s^{-2}}\right)\right)$$

$$\geq \frac{s^m}{2}\exp\left(\frac{m}{2}\sqrt{1-s^{-2}}\right),$$

by the inequality $\log(1+t) \geq \frac{t}{2}$, $t \in [0,1]$. Substituting in (11.50) gives for $x \in [0,1)$,

$$|f(x) - S_{2m}(x)|$$

$$\leq C\int_1^\infty \frac{(s-1)^{-3/4}}{(s-x)\, s^m} e^{-m(1-s^{-1})^{1/2}}\, ds$$

$$\leq \frac{C}{1-x}\left[\int_1^2 (s-1)^{-3/4}\, e^{-\frac{m}{2}(s-1)^{1/2}}\, ds + e^{-Cm}\int_2^\infty \frac{ds}{s^m}\right]$$

$$\leq \frac{C}{1-x}\left[\frac{2}{\sqrt{m}}\int_0^{m^2} t^{-3/4} e^{-\frac{1}{2}t^{1/2}}\, dt + e^{-Cm}\frac{2^{-m+1}}{m-1}\right] \leq \frac{C}{\sqrt{m}\,(1-x)}.$$

Then for $x \in [-1,1]$, we have

$$\left|(1-x^2)^{1/4} - (1-x^2)\, S_{2m}(x)\right| \leq \frac{C}{\sqrt{m}},$$

so also for $|x| \leq 1 - m^{-2+\varepsilon}$,

$$\left|1 - \frac{(1-x^2)\, S_{2m}(x)}{(1-x^2)^{1/4}}\right| \leq \frac{C}{[m^2(1-x^2)]^{1/4}} \leq Cm^{-\varepsilon/4},$$

and hence also for such $x$,

$$\left|1 - \frac{R_m(x)}{(1-x^2)^{1/2}}\right| \leq Cm^{-\varepsilon/4}. \tag{11.51}$$

Let us summarize what we have so far: $R_m$ has degree $\leq 4m$, and satisfies (11.49) and (11.51). We need a better lower bound than that in (11.49). So set

$$Y_m(x) = \frac{R_m(x) + (1 - x^2) m^{-\varepsilon/4}}{1 + m^{-\varepsilon/4}}.$$

Then (11.44) follows easily from the definition of $Y_m$ and (11.49). Likewise the definition of $Y_m$ and (11.51) give (11.45). Finally, Bernstein's inequality gives

$$|Y_m'(x)| \leq \frac{4m}{\sqrt{1 - x^2}} \|Y_m\|_{L_\infty[-1,1]} \leq \frac{4m}{\sqrt{1 - x^2}}$$

and then (11.46) also follows.

(II) Here we choose $f(t) = (1 - t^2)^{1/2}$, and again let $S_{2m}(t)$ be the polynomial of degree $\leq 2m - 2$ that interpolates to $f$ at the zeros of $T_m^2$. Since for $m \geq 1, f^{(2m)} \leq 0$ in $(-1, 1)$, (11.48) gives

$$S_{2m}(t) \geq (1 - t^2)^{1/2}, \quad t \in (-1, 1). \tag{11.52}$$

We set $Y_m = S_{2m}$. By using the error formula for Lagrange interpolation much as in (I), we obtain for $x \in [0, 1)$,

$$|f(x) - S_{2m}(x)| \leq C \int_1^\infty \frac{(s - 1)^{1/2}}{(s - x) s^m} e^{-\frac{m}{2}(1 - s^{-1})^{1/2}} ds \leq \frac{C}{m^3 (1 - x)}.$$

Thus for $x \in (-1, 1)$,

$$\left| 1 - \frac{S_{2m}(x)}{(1 - x^2)^{1/2}} \right| \leq \frac{C}{[m^2 (1 - x^2)]^{3/2}}. \tag{11.53}$$

Then (11.47) follows from this last inequality and (11.52). Next, this last inequality gives something stronger than (11.45), and the Markov-Bernstein inequality gives (11.46) much as in (I). □

*Proof of Theorem* 11.1. (b) We let $\eta > 0$ be a small positive number and

$$S_n(t) = \left\{ \frac{|X_n(t)|^2 + \eta n^{-\tau_1} E_n(t)}{1 + n^{-\tau_1}} \right\} Y_k(t),$$

where $X_n, E_n$ are as in Lemmas 11.6, 11.7 and $Y_k$ is as in Lemma 11.8 (I), and $k = k(n) = \left[ \frac{1}{4} n^{1/3 - 2\tau_1} \right]$. Recall also that $\iota_1 \in (0, \frac{\alpha}{50}]$. We also set in Lemma 11.8,

$$\varepsilon = 24\tau_1 \in \left( 0, \frac{1}{2} \right).$$

Observe that

$$k^\varepsilon \sim n^{8\tau_1 - 48\tau_1^2} \geq n^{7\tau_1}. \tag{11.54}$$

Then $S_n$ has degree $\leq 2n - n^{1/3-2\tau_1}$, and for $t \in [-1, 1]$, by (11.34), (11.43), and (11.44),

$$\psi_n(t) = \left\{ S_n(t) e^{-2nQ_n(t)} \left(1 - t^2\right)^{-1/2} \right\}^{1/2}$$

$$= \left\{ \left[ \frac{|X_n(t)|^2 + \eta n^{-\tau_1} E_n(t)}{1 + n^{-\tau_1}} \right] e^{-2nQ_n(t)} \left[ Y_k(t) \left(1 - t^2\right)^{-1/2} \right] \right\}^{1/2}.$$

$$\leq \left\{ \frac{1 + C_2 \eta n^{-\tau_1}}{1 + n^{-\tau_1}} 1 \right\}^{1/2} \leq 1,$$

if $\eta \leq 1/C_2$. Thus we have the upper bound in (11.4). Also, for $t \in [-1, 1]$, (11.43) and (11.44) give

$$\psi_n(t) \geq \left\{ \frac{\eta n^{-\tau_1} E_n(t) e^{-2nQ_n(t)}}{1 + n^{-\tau_1}} \frac{\sqrt{1 - t^2}}{2k^{\varepsilon/4}} \right\}^{1/2} \geq C \left\{ n^{-3\tau_1} \sqrt{1 - t^2} \right\}^{1/2},$$

so we also have the lower bound in (11.4).

(a) Next, if both $|t| \leq 1 - 4n^{-\frac{2}{3}+8\tau_1}$, and $|t| \leq 1 - k^{-2+\varepsilon}$,

$$\psi_n(t) = \left\{ \left[ \frac{|X_n(t)|^2 + \eta n^{-\tau_1} E_n(t)}{1 + n^{-\tau_1}} \right] e^{-2nQ_n(t)} \left[ Y_k(t) \left|1 - t^2\right|^{-1/2} \right] \right\}^{1/2}$$

$$= \left\{ \frac{1 + O\left(n^{-\tau_1}\right) + O\left(n^{-\tau_1}\right)}{1 + n^{-\tau_1}} \right\}^{1/2} \left(1 + O\left(k^{-\varepsilon/4}\right)\right) = 1 + O\left(n^{-\tau_1}\right),$$

by (11.36), (11.43), and (11.45) and since $k^{-\varepsilon/4} \ll n^{-\tau_1}$. It is easily seen that for large enough $n$, the range $|t| \leq 1 - 4n^{-\frac{2}{3}+8\tau_1}$ is larger than $|t| \leq 1 - k^{-2+\varepsilon}$. Indeed $k^{-2+\varepsilon} \sim n^{-\frac{2}{3}+12\tau_1 - 48\tau_1^2} \geq n^{-\frac{2}{3}+11\tau_1}$. So we have (11.3) for $|t| \leq 1 - k^{-2+\varepsilon}$ and hence for $|t| \leq 1 - 4n^{-2/3+12\tau_1}$.

(c) Note first that by (11.43) and Theorem 4.1,

$$\left\| E_n e^{-2nQ_n} \right\|_{L_\infty(I_n)} \leq C_2,$$

and then (8.1) of Theorem 8.1 gives for $x \in [-1, 1]$,

$$\left| \frac{d}{dx} \left( E_n(x) e^{-2nQ_n(x)} \right) \right| \leq Cn.$$

Then using (11.3) and (11.36), (11.37), (11.45), (11.46), for $t$ satisfying $|t| \leq 1 - 4n^{-2/3+8\tau_1}$, and $|t| \leq 1 - k^{-2+\varepsilon}$,

$$\frac{d}{dt} \log \psi_n (t)$$

$$= \frac{1}{2} \frac{d}{dt} \log \left\{ \left[ |X_n (t)|^2 + \eta n^{-\tau_1} E_n (t) \right] e^{-2nQ_n(t)} \right\}$$

$$+ \frac{1}{2} \frac{d}{dt} \left\{ \log \left[ Y_k (t) \left| 1 - t^2 \right|^{-1/2} \right] \right\}$$

$$= O \left( \frac{d}{dt} \left\{ |X_n (t)|^2 e^{-2nQ_n(t)} \right\} \right) + O \left( \frac{d}{dt} \left\{ \eta n^{-\tau_1} E_n (t) e^{-2nQ_n(t)} \right\} \right)$$

$$+ O \left( | Y_k' (t) | \right) + O \left( (1 - t^2)^{-1} \right)$$

$$= O \left( n^{1-\tau_1} \right) + O \left( n^{1-\tau_1} \right) + O \left( k^{2-\varepsilon/2} \right) + O \left( k^{2-\varepsilon} \right) = O \left( n^{1-\tau_1} \right).$$

Then (11.5) follows, recall that the two ranges of $t$ both contain the range $|t| \leq 1 - 4n^{-2/3+12\tau_1}$.

(d) Now our upper and lower bounds for $\psi_n$ give

$$Cn^{-4\tau_1} \left( 1 - t^2 \right) \leq \left| S_n (t) e^{-2nQ_n(t)} \right| \leq 1, \quad t \in [-1, 1], \tag{11.55}$$

so our Markov-Bernstein inequality (8.1) gives

$$\left| \frac{d}{dt} \left( S_n (t) e^{-2nQ_n(t)} \right) \right| \leq Cn \left( 1 - t^2 + n^{-2/3} \right)^{1/2}, \quad t \in [-1, 1]. \tag{11.56}$$

Next,

$$\frac{\psi_n' (t)}{\psi_n (t)} = \frac{t}{2 (1 - t^2)} + \frac{1}{2} \frac{\frac{d}{dt} \left( S_n (t) e^{-2nQ_n(t)} \right)}{S_n (t) e^{-2nQ_n(t)}}$$

$$= \frac{t}{2 (1 - t^2)} + \frac{1}{2} \frac{\frac{d}{dt} \left( S_n (t) e^{-2nQ_n(t)} \right)}{(1 - t^2)^{1/2} \psi_n (t)^2}, \tag{11.57}$$

so combining (11.56), (11.4) gives

$$\left| \frac{\psi_n' (t)}{\psi_n (t)} \right| \leq \frac{C}{1 - t^2} + C \frac{n^{1+4\tau_1} \left( 1 - t^2 + n^{-2/3} \right)^{1/2}}{1 - t^2}$$

$$\leq \frac{C}{1 - t^2} + C \frac{n^{1+4\tau_1}}{(1 - t^2)^{1/2}} + C \frac{n^{2/3+4\tau_1}}{1 - t^2}.$$

Then the result follows.

(e) Now from (11.3),

$$\int_{|t|\leq 1-4n^{-2/3+12\tau_1}} \frac{|\log \psi_n(t)|}{\sqrt{1-t^2}} dt = O(n^{-\tau_1}).$$

Next, our bounds in (11.4) give

$$\int_{1-4n^{-2/3+12\tau_1}\leq |t|\leq 1} \frac{|\log \psi_n(t)|}{\sqrt{1-t^2}} dt$$

$$= O\left((\log n)\, n^{-1/3+6\tau_1}\right) + O\left(\int_{1-4n^{-2/3+12\tau_1}\leq |t|\leq 1} \frac{|\log(1-t^2)|}{\sqrt{1-t^2}} dt\right)$$

$$= O\left((\log n)\, n^{-1/3+6\tau_1}\right) = O(n^{-\tau_1}).$$

Combining the two estimates gives (11.7).

(f) Let $|x| \leq 1 - n^{-\tau_1}$. We split

$$\int_{-1}^{1} \frac{\left|1 - \left(\frac{\psi_n(t)}{\psi_n(x)}\right)^2\right| \psi_n(t)^{-1}}{|x-t|} \frac{dt}{\sqrt{1-t^2}} = \left(\int_{\mathcal{I}_1} + \int_{\mathcal{I}_2} + \int_{\mathcal{I}_3}\right) \cdots$$

$$=: J_1 + J_2 + J_3.$$

Here the ranges are:

$$\mathcal{I}_1 : |t| \leq 1 - 4n^{-\frac{2}{3}+12\tau_1}, |x-t| \geq n^{-6};$$

$$\mathcal{I}_2 : |t| \leq 1 - 4n^{-\frac{2}{3}+12\tau_1}, |x-t| < n^{-6};$$

$$\mathcal{I}_3 : 1 - 4n^{-\frac{2}{3}+12\tau_1} \leq |t| < 1.$$

Here using (11.3) and assuming $x \geq 0$,

$$J_1 \leq Cn^{-\tau_1} \int_{\mathcal{I}_1} \frac{1}{|x-t|} \frac{dt}{\sqrt{1-t}}$$

$$\leq Cn^{-\tau_1} (1-x)^{-1/2} \int_{\mathcal{J}_1} \frac{1}{|s-1|} \frac{ds}{\sqrt{s}},$$

where we have made the substitution $1-t = (1-x)s$, and where

$$\mathcal{J}_1 = \left\{ s \in (0,\infty) : |s-1| \geq \frac{1}{n^6(1-x)} \right\}.$$

Then we can continue this as

$$J_1 \leq Cn^{-\tau_1} (1-x)^{-1/2} \log n \leq Cn^{-\tau_1/3},$$

as $1 - x \geq n^{-\tau_1}$. Next, using (11.5), and $\psi_n \leq 1$,

$$J_2 \leq C \sup_{t \in \mathcal{I}_2} \left| \frac{d}{dt} \psi_n^2 (t) \right| \left| \int_{\mathcal{I}_2} \frac{dt}{\sqrt{1-t}} \right| \leq Cn^{1-\tau_1} n^{-3} \leq Cn^{-\tau_1}.$$

Finally,

$$J_3 \leq \frac{C}{1 - |x|} \int_{1-4n^{-\frac{2}{3}+12\tau_1} \leq |t| \leq 1} \left( \psi_n (t) + \psi_n (t)^{-1} \right) \frac{dt}{\sqrt{1-t^2}}$$

$$\leq Cn^{3\tau_1} \int_{1-4n^{-\frac{2}{3}+12\tau_1}}^{1} \frac{dt}{(1-t)^{3/4}}$$

$$\leq Cn^{-\frac{1}{6}+6\tau_1} \leq Cn^{-\tau_1}.$$

Adding the estimates gives (11.8).

(g) This is similar to, but easier than (f).

(h) Let $1 > x \geq 1 - n^{-\tau_1}$. (The case of negative $x$ is similar.) We split

$$\int_{-1}^{1} \frac{\left| 1 - \left( \frac{\psi_n(t)}{\psi_n(x)} \right)^2 \right| \psi_n (t)^{-1}}{|x - t|} \frac{dt}{\sqrt{1-t^2}} = \left( \int_{\mathcal{I}_1} + \int_{\mathcal{I}_2} \right) \cdots =: J_1 + J_2.$$

Here the ranges are:

$$\mathcal{I}_1 : t \in (-1, 1), \quad |x - t| \geq n^{-6} (1 - |x|)^2 ;$$

$$\mathcal{I}_2 : t \in (-1, 1), \quad |x - t| < n^{-6} (1 - |x|)^2 .$$

Some of these ranges may be empty. Using our bounds (11.4), and then the substitution $1 - t = s (1 - x)$,

$$J_1 = \int_{\mathcal{I}_1} \frac{\left| 1 - \left( \frac{\psi_n(t)}{\psi_n(x)} \right)^2 \right| \psi_n (t)^{-1}}{|x - t|} \frac{dt}{\sqrt{1-t^2}}$$

$$\leq C \int_{\mathcal{I}_1} \frac{\psi_n (t)^{-1} + \psi_n (t) \psi_n (x)^{-2}}{|x - t|} \frac{dt}{\sqrt{1-t^2}}$$

$$\leq C \int_{\mathcal{I}_1} \frac{n^{2\tau_1} (1 - t^2)^{-1/4} + n^{4\tau_1} (1 - x^2)^{-1/2}}{|x - t|} \frac{dt}{\sqrt{1-t^2}}$$

$$\leq Cn^{2\tau_1} \int_{\mathcal{I}_1} \frac{1}{|x - t|} \frac{dt}{(1-t)^{3/4}} + Cn^{4\tau_1} (1 - x^2)^{-1/2} \int_{\mathcal{I}_1} \frac{1}{|x - t|} \frac{dt}{(1-t)^{1/2}}$$

$$\leq C n^{2\tau_1} (1-x)^{-3/4} \int_{|s-1|\geq n^{-6}/(1-x)} \frac{1}{|s-1|} \frac{ds}{|s|^{3/4}}$$

$$+ C n^{4\tau_1} (1-x)^{-1} \int_{|s-1|\geq n^{-6}/(1-x)} \frac{1}{|s-1|} \frac{ds}{|s|^{1/2}}$$

$$\leq C n^{2\tau_1} (1-x)^{-3/4} \log\left[ n^6 (1-x)^{-1} \right] + C n^{4\tau_1} (1-x)^{-1} \log\left[ n^6 (1-x)^{-1} \right]$$

$$\leq C n^{5\tau_1} (1-|x|)^{-5/4}.$$

Next, for $|x-t| < n^{-6} (1-x)^2$, we have $1 - t \sim 1 - x$, and for some $\xi$ between $x, t$,

$$\frac{\left| 1 - \left( \frac{\psi_n(t)}{\psi_n(x)} \right)^2 \right| \psi_n(t)^{-1}}{|x-t|}$$

$$= \psi_n(x)^{-2} \, 2\psi_n(\xi) \, \psi_n'(\xi) \, \psi_n(t)^{-1}$$

$$= 2\psi_n(x)^{-2} \, \psi_n(t)^{-1} \, \psi_n(\xi)^2 \, \frac{\psi_n'(\xi)}{\psi_n(\xi)}$$

$$\leq C n^{6\tau_1} \left( 1 - x^2 \right)^{-3/4} \left\{ \frac{n^{1+4\tau_1}}{(1-x^2)^{1/2}} + C \frac{n^{2/3+4\tau_1}}{1-x^2} \right\}$$

$$\leq C n^{1+10\tau_1} \left( 1-x^2 \right)^{-5/4} + C n^{2/3+10\tau_1} \left( 1-x^2 \right)^{-7/4},$$

by (11.4), (11.6). Then

$$J_2 \leq C \left( n^{1+10\tau_1} \left( 1-x^2 \right)^{-5/4} + n^{2/3+10\tau_1} \left( 1-x^2 \right)^{-7/4} \right) \int_{\mathcal{I}_2} \frac{dt}{\sqrt{1-t}}$$

$$\leq C \left( n^{1+10\tau_1} \left( 1-x^2 \right)^{-5/4} + n^{2/3+10\tau_1} \left( 1-x^2 \right)^{-7/4} \right) n^{-6} \left( 1-x^2 \right)^{3/2}$$

$$\leq C(1-x^2)^{-1/4}.$$

Together with our estimate for $J_1$, this gives the result.

(i) We split

$$\int_{-1}^{1} \left| \frac{\frac{\psi_n'(t)}{\psi_n(t)} (1-t^2) - \frac{\psi_n'(x)}{\psi_n(x)} (1-x^2)}{t-x} \right| \frac{dt}{\sqrt{1-t^2}} = \left( \int_{\mathcal{I}_1} + \int_{\mathcal{I}_2} + \int_{\mathcal{I}_3} \right) \cdots$$

$$=: J_1 + J_2 + J_3,$$

where the ranges are:

$$\mathcal{I}_1 : |t| \leq 1 - 4n^{-\frac{2}{3}+12\tau_1}, \quad |x-t| \geq n^{-100};$$

$$\mathcal{I}_2 : |t| \leq 1 - 4n^{-\frac{2}{3}+12\tau_1}, \quad |x-t| < n^{-100};$$

$$\mathcal{I}_3 : 1 - 4n^{-\frac{2}{3}+12\tau_1} \leq |t| < 1.$$

Using our estimates (11.3) and (11.5),

$$J_1 \leq C \int_{\mathcal{I}_1} \frac{n^{1-\tau_1}}{|t-x|} \frac{dt}{\sqrt{1-t^2}}$$

$$\leq Cn^{1-\tau_1} (\log n) \left(1 - x^2\right)^{-1/2}. \tag{11.58}$$

Next, from (11.5) and (11.6),

$$J_3 \leq C \int_{\mathcal{I}_3} \frac{n^{1+4\tau_1} \sqrt{1-t^2} + n^{2/3+4\tau_1} + n^{1-\tau_1}}{|t-x|} \frac{dt}{\sqrt{1-t^2}}$$

$$\leq Cn^{1+4\tau_1} \left(1 - 4n^{-\frac{2}{3}+12\tau_1} - |x|\right)^{-1} n^{-\frac{2}{3}+12\tau_1}$$

$$+ Cn^{1-\tau_1} \left(1 - 4n^{-\frac{2}{3}+12\tau_1} - |x|\right)^{-1} n^{-\frac{1}{3}+6\tau_1}$$

$$\leq Cn^{\frac{1}{3}+17\tau_1} + Cn^{\frac{2}{3}+6\tau_1} \leq Cn^{\frac{2}{3}+6\tau_1}. \tag{11.59}$$

Next, we estimate the central term $J_2$. This is messy as we do not know that $Q_n''$ exists, and have to use the Lipschitz $\alpha$ condition on $Q_n'$. In this range, $1-|t| \sim 1-|x|$. Observe that from (11.57),

$$\Delta(x,t) = \left| \frac{\psi_n'(t)}{\psi_n(t)} \left(1-t^2\right) - \frac{\psi_n'(x)}{\psi_n(x)} \left(1-x^2\right) \right|$$

$$= \frac{1}{2} \left| t - x + \frac{(1-t^2)\frac{d}{dt} \left(S_n(t) e^{-2nQ_n(t)}\right)}{S_n(t) e^{-2nQ_n(t)}} \right.$$

$$\left. - \frac{(1-x^2)\frac{d}{dx} \left(S_n(x) e^{-2nQ_n(x)}\right)}{S_n(x) e^{-2nQ_n(x)}} \right|$$

$$\leq \frac{1}{2} |t-x| + |x^2 - t^2| \left| \frac{\frac{d}{dt} \left(S_n(t) e^{-2nQ_n(t)}\right)}{S_n(t) e^{-2nQ_n(t)}} \right|$$

$$+ \left(1-x^2\right) \left| \frac{\frac{d}{dt} \left(S_n(t) e^{-2nQ_n(t)}\right)}{S_n(t) e^{-2nQ_n(t)}} - \frac{\frac{d}{dx} \left(S_n(x) e^{-2nQ_n(x)}\right)}{S_n(x) e^{-2nQ_n(x)}} \right|$$

$$\leq C|t-x| \frac{n^{1+4\tau_1} \left(1 - t^2 + n^{-2/3}\right)^{1/2}}{1-t^2} + \left(1-x^2\right) |T|, \tag{11.60}$$

where we have used (11.55) and (11.56), and

$$
T = \left| \frac{\frac{d}{dt}\left(S_n(t)\,e^{-2nQ_n(t)}\right)}{S_n(t)\,e^{-2nQ_n(t)}} - \frac{\frac{d}{dx}\left(S_n(x)\,e^{-2nQ_n(x)}\right)}{S_n(x)\,e^{-2nQ_n(x)}} \right|
$$

$$
\le \left| \frac{S_n'(t)}{S_n(t)} - \frac{S_n'(x)}{S_n(x)} \right| + 2n\left| Q_n'(t) - Q_n'(x) \right|
$$

$$
\le \left| \frac{S_n'(t)\,e^{-nQ_n(t)} - S_n'(x)\,e^{-nQ_n(x)}}{S_n(t)\,e^{-nQ_n(t)}} \right|
$$

$$
+ \left| S_n'(x)\,e^{-nQ_n(x)} \right| \left| \frac{1}{S_n(t)\,e^{-nQ_n(t)}} - \frac{1}{S_n(x)\,e^{-nQ_n(x)}} \right|
$$

$$
+ 2n\left| Q_n'(t) - Q_n'(x) \right|
$$

$$
\le Cn^{2+4\tau_1}\left(1-x^2\right)^{-1}|t-x| + Cn^{2+8\tau_1}\left(1-x^2\right)^{-2}|t-x| + Cn\,|t-x|^\alpha,
$$

by (1.11), (11.55), (11.56), and two applications of Theorem 8.1. Substituting this into (11.60) gives

$$
\Delta(x,t) \le Cn^{2+8\tau_1}\left(1-x^2\right)^{-1}|t-x| + Cn\,|t-x|^\alpha.
$$

Then

$$
J_2 = \int_{\mathcal{I}_2} \left| \frac{\Delta(x,t)}{t-x} \right| \frac{dt}{\sqrt{1-t^2}}
$$

$$
\le Cn^{2+8\tau_1}\left(1-x^2\right)^{-1}\int_{\mathcal{I}_2} \frac{dt}{\sqrt{1-t^2}} + Cn\int_{\mathcal{I}_2} |t-x|^{\alpha-1}\frac{dt}{\sqrt{1-t^2}}
$$

$$
\le Cn^{-98+8\tau_1}\left(1-x^2\right)^{-3/2} + Cn^{1+4\tau_1}n^{-100\alpha} \le Cn^{1-\tau_1},
$$

by our restrictions on $x$ and $\tau_1$. Together with (11.58) and (11.59), this gives

$$
\int_{-1}^{1} \left| \frac{\frac{\psi_n'(t)}{\psi_n(t)}\left(1-t^2\right) - \frac{\psi_n'(x)}{\psi_n(x)}\left(1-x^2\right)}{t-x} \right| \frac{dt}{\sqrt{1-t^2}}
$$

$$
\le Cn^{1-\tau_1}(\log n)\left(1-x^2\right)^{-1/2} + Cn^{2/3+6\tau_1} + Cn^{1-\tau_1} \le Cn^{1-\tau_1/3},
$$

if $|x| \le 1 - n^{-\tau_1}$.                                                                    □

*Proof of Theorem* 11.2. (a), (b) We let

$$
S_n(t) = \left\{ |X_n(\rho_n t)|^2 \right\} Y_k(t),
$$

where $X_n$ and $Y_k$ are as in Lemmas 11.6 and 11.8(I) respectively, and $k = k(n) = \left[\frac{1}{4}n^{1/3-2\tau_1}\right]$. We also set, as in the previous proof,

$$\varepsilon = 24\tau_1 \in \left(0, \frac{1}{2}\right).$$

Then $S_n$ has degree $\leq 2n - n^{1/3-2\tau_1}$, and for $t \in [-1, 1]$,

$$\hat{\psi}_n(t) = |X_n(\rho_n t)|^2 e^{-2nQ_n(\rho_n t)} \left[ Y_k(t) \left|1 - t^2\right|^{-1/2} \right]^{1/2} \leq 1,$$

by (11.34) and (11.44). So we have (11.15). Also from (11.36) and (11.45), we have

$$\hat{\psi}_n(t) = 1 + O\left(n^{-\tau_1}\right) + O\left(k^{-\varepsilon/4}\right) = 1 + O\left(n^{-\tau_1}\right),$$

as in the proof of Theorem 11.1. This holds for $t$ satisfying $|t| \leq 1 - k^{-2+\varepsilon}$ such that also

$$|\rho_n t| \leq 1 - 4n^{-\frac{2}{3}+8\tau_1}$$

$$\Leftrightarrow |t| \leq \frac{1 - 4n^{-\frac{2}{3}+8\tau_1}}{1 + \left(A\frac{\log n}{n}\right)^{2/3}}$$

$$= 1 - 4n^{-\frac{2}{3}+8\tau_1}(1 + o(1)).$$

Since $k^{-2+\varepsilon} \sim n^{-2/3+12\tau_1-48\tau_1^2}$, the first requirement is more severe. So it suffices that $|t| \leq 1 - n^{-\frac{2}{3}+12\tau_1}$. Then we have (11.14). Also,

$$\int_{-1+n^{-\frac{2}{3}+12\tau_1}}^{1-n^{-\frac{2}{3}+12\tau_1}} \frac{\left|\log \hat{\psi}_n(t)\right|}{\sqrt{1-t^2}} dt \leq Cn^{-\tau_1}.$$

Next, for $1 \geq |t| \geq 1 - n^{-\frac{2}{3}+12\tau_1}$, we have $|\rho_n t| \leq 1 + \left(A\frac{\log n}{n}\right)^{2/3}$, so our lower bound (11.35) (with $\Delta$ very small) gives

$$|X_n(\rho_n t)|^2 e^{-2nQ_n(\rho_n t)} \geq \exp\left(-n^{3\tau_1}\right).$$

Then from (11.47),

$$\hat{\psi}_n(t) \geq \exp\left(-Cn^{3\tau_1}\right)(1 - t^2)^{1/4},$$

so

$$\int_{1 \geq |t| \geq 1-n^{-\frac{2}{3}+12\tau_1}} \frac{\left|\log \hat{\psi}_n(t)\right|}{\sqrt{1-t^2}} dt$$

$$= -\int_{1 \geq |t| \geq 1-n^{-\frac{2}{3}+12\tau_1}} \frac{\log \hat{\psi}_n(t)}{\sqrt{1-t^2}} dt$$

$$\leq C \int_{1 \geq |t| \geq 1-n^{-\frac{2}{3}+12\tau_1}} \frac{n^{3\tau_1} + \left|\log\left(1-t^2\right)\right|}{\sqrt{1-t^2}} dt$$

$$\leq n^{-\tau_1},$$

for large enough $n$. So we have (11.16). □

For the proof of Theorem 11.3, we need polynomials that are small in most of $[-1, 1]$, but close to 1 near $\pm 1$:

**Lemma 11.9.** *There exist polynomials $Z_m$ of degree $\leq m$ such that*

$$0 < Z_m < 1 \text{ in } [-1, 1]; \tag{11.61}$$

$$0 < Z_m \leq \frac{C_1}{m^{1/2}}, \quad |x| \leq 1 - 2m^{-1/2}; \tag{11.62}$$

$$0 \leq 1 - Z_m(x) \leq \frac{C_1}{m^{1/2}}, \quad 1 - m^{-1/2} \leq |x| \leq 1. \tag{11.63}$$

*Proof.* Let $\varepsilon = m^{-1/2}$ and

$$f(x) = \begin{cases} 0, & |x| \leq 1 - 2\varepsilon; \\ 1, & 1 - \varepsilon \leq |x| \leq 1. \end{cases}$$

We may define $f$ in $1 - \varepsilon \leq |x| \leq 1 - 2\varepsilon$ so that $f$ is continuously differentiable in $[-1, 1]$ with

$$\left|f'(x)\right| \leq \frac{2}{\varepsilon}, \quad x \in [-1, 1].$$

(Smooth a piecewise linear function near $\pm(1-\varepsilon)$ and $\pm(1-2\varepsilon)$.) By Jackson's Theorem, we may find polynomials $Z_m$ of degree $\leq m$ such that

$$\|f - Z_m\|_{L_\infty[-1,1]} \leq \frac{C}{m\varepsilon} = \frac{C}{m^{1/2}}.$$

We can assume $0 < Z_m < 1$ in $[-1, 1]$, by adding a small positive constant and then dividing by $1 + Am^{-1/2}$ for large enough $A$. We easily obtain (11.62–3) from the last inequality. □

*Proof of Theorem* 11.3. We let $B$ be a large enough positive number, and

$$S_n(t) = \left\{ |X_n(t)|^2 + BE_n(t) Z_m(t) \right\} Y_k(t),$$

where $X_n$ and $Y_k$ are as in Lemmas 11.6 and 11.8(II) respectively, and $k = \left[ \frac{1}{4} n^{1/3 - 2\tau_1} \right]$. Moreover, $Z_m$ is as in Lemma 11.9 with $m \sim n^{2\tau_1}$. Then $E_n Z_m$ has degree $\leq 2n - 3n^{1/3} + n^{2\tau_1} \leq 2n - 2n^{1/3}$ for large enough $n$, so $S_n$ has degree $\leq 2n - n^{1/3 - 2\tau_1}$. By (11.36), (11.45), and (11.62),

$$\psi_n(t) = 1 + O(n^{-\tau_1}) + O(n^{-\tau_1}) + O(k^{-\varepsilon/4}),$$

for $t$ satisfying all of $|t| \leq 1 - 4n^{-2/3 + 8\tau_1}$ and $|t| \leq 1 - n^{-\tau_1}$ and $|t| \leq 1 - k^{-2 + \varepsilon}$. As in the proof of Theorem 11.1, we see that then

$$\psi_n(t) = 1 + O(n^{-\tau_1}), \tag{11.64}$$

for $|t| \leq 1 - n^{-\tau_1}$. Also, for $1 - n^{-\tau_1} \leq |t| \leq 1 - 4n^{-2/3 + 8\tau_1}$, we have by (11.36) and (11.47),

$$\psi_n^2(t) \geq \left\{ |X_n(t)|^2 e^{-2nQ_n(t)} \right\} Y_k(t) \left( 1 - t^2 \right)^{-1/2} \geq 1 + O(n^{-\tau_1})$$

while for $1 - n^{-\tau_1} \leq |t| \leq 1$, by (11.43), (11.47), (11.63),

$$\psi_n^2(t) \geq \left\{ BE_n(t) e^{-2nQ_n(t)} Z_m(t) \right\} Y_k(t) \left( 1 - t^2 \right)^{-1/2}$$

$$\geq BC \left( 1 - \frac{C_1}{n^{\tau_1}} \right) \geq 1,$$

provided $B$ is large enough. Since $1 - n^{-\tau_1} < 1 - 4n^{-2/3 + 8\tau_1}$, we have proved that for all $t \in [-1, 1]$

$$\psi_n^2(t) \geq 1 + O(n^{-\tau_1}).$$

Multiplying $S_n$ by $1 + Dn^{-\tau_1}$ with $D$ large enough gives (11.17). Next, (11.64) shows that

$$\int_{-1 + n^{-\tau_1}}^{1 - n^{-\tau_1}} \frac{|\log \psi_n(t)|}{\sqrt{1 - t^2}} dt \leq Cn^{-\tau_1}.$$

Finally, using (11.34), (11.43), (11.47), and (11.61), we see that

$$\int_{[-1,1] \setminus [-1 + n^{-\tau_1}, 1 - n^{-\tau_1}]} \frac{|\log \psi_n(t)|}{\sqrt{1 - t^2}} dt \leq C (\log n) n^{-\tau_1/2}.$$

Then (11.18) follows.                                                            □

# Chapter 12
# Formulae Involving Bernstein-Szegő Polynomials

In this chapter, we state some formulae involving orthonormal polynomials for Bernstein-Szegő weights $w_{n,B}^2$, where

$$w_{n,B}(t) = \frac{\left(1 - t^2\right)^{1/4}}{\sqrt{S_n(t)}}.$$

Here $S_n$ is a polynomial of degree $2q_n \leq 2n$, that is positive in $[-1, 1]$, except possibly for simple zeros at $\pm 1$. In the previous chapter, we approximated $e^{-2nQ_n}$ by such a weight, and in subsequent chapters, we shall use formulae for $p_n\left(w_{n,B}^2, x\right) = \gamma_n\left(w_{n,B}^2\right) x^n + \cdots$ and related quantities, to obtain asymptotics for quantities associated with $e^{-2nQ_n}$. We need some more notation. Throughout we assume that $\{Q_n\} \in \mathcal{Q}$. For a function $g : [-1, 1] \rightarrow \mathbb{R}$, we let $\breve{g}(\theta) = g(\cos\theta)$, and in particular, we let

$$\breve{w}_{n,B}(\theta) = w_{n,B}(\cos\theta) \tag{12.1}$$

and

$$f_n(\theta) = w_{n,B}^2(\cos\theta) |\sin\theta| = \frac{\sin^2\theta}{S_n(\cos\theta)}. \tag{12.2}$$

Define the associated *Szegő function*

$$D(f_n; z) = \exp\left(\frac{1}{4\pi} \int_{-\pi}^{\pi} \frac{e^{it} + z}{e^{it} - z} \log f_n(t)\, dt\right), \quad |z| < 1, \tag{12.3}$$

© The Author(s) 2018
E. Levin, D.S. Lubinsky, *Bounds and Asymptotics for Orthogonal Polynomials for Varying Weights*, SpringerBriefs in Mathematics,
https://doi.org/10.1007/978-3-319-72947-3_12

that satisfies

$$\left|D\left(f_n; e^{it}\right)\right|^2 = f_n(t), \quad t \in (-\pi, \pi) \setminus \{0\}. \tag{12.4}$$

The argument of the Szegő function on the unit circle is [25, p. 420]

$$\Gamma(f_n; \theta) = \arg D\left(f_n; e^{i\theta}\right) = \frac{1}{4\pi} PV \int_{-\pi}^{\pi} \left(\cot \frac{\theta - t}{2}\right) \log f_n(t)\, dt$$

$$= \frac{1}{4\pi} \int_{-\pi}^{\pi} \left(\cot \frac{\theta - t}{2}\right) [\log f_n(t) - \log f_n(\theta)]\, dt. \tag{12.5}$$

Here $PV$ denotes Cauchy Principal Value. If $x \in (-1, 1)$ and $x = \cos\theta$, where $\theta \in (0, \pi)$, then also [25, eqn. (21), p. 441; eqn. (41), p. 433]

$$\Gamma(f_n; \theta) = \Gamma\left(\breve{w}_{n,B}^2; \theta\right) + \Gamma(|\sin \cdot|; \theta)$$

$$= \frac{\sqrt{1 - x^2}}{2\pi} \int_{-1}^{1} \frac{\log w_{n,B}^2(s) - \log w_{n,B}^2(x)}{s - x} \frac{ds}{\sqrt{1 - s^2}} + \frac{\theta}{2} - \frac{\pi}{4}. \tag{12.6}$$

Also, its derivative satisfies [25, eqn. (23), p. 441]

$$\Gamma'(f_n; \theta) = \Gamma'\left(\breve{w}_{n,B}^2; \theta\right) + \frac{1}{2}$$

$$= -\frac{1}{2\pi} PV \int_{-1}^{1} \left\{\frac{d}{ds}\left(\log w_{n,B}^2(s)\right)\right\} \frac{\sqrt{1 - s^2}}{s - x}\, ds + \frac{1}{2} \tag{12.7}$$

$$= -\frac{1}{\pi} \int_{-1}^{1} \frac{\left(w_{n,B}'(s)/w_{n,B}(s)\right)\left(1 - s^2\right) - \left(w_{n,B}'(x)/w_{n,B}(x)\right)\left(1 - x^2\right)}{s - x}$$

$$\times \frac{ds}{\sqrt{1 - s^2}} + \frac{1}{2}. \tag{12.8}$$

We can now state:

**Lemma 12.1.** Let $m \geq q_n = \frac{1}{2} \deg(S_n)$. Let $x = \cos\theta$, where $\theta \in [0, \pi]$, and let $z = e^{i\theta}$.
(a)

$$\gamma_m\left(w_{n,B}^2\right) = \frac{2^m}{\sqrt{2\pi}} \exp\left(-\frac{1}{4\pi} \int_{-\pi}^{\pi} \log f_n(t)\, dt\right)$$

$$= \frac{2^m}{\sqrt{2\pi}} \exp\left(-\frac{1}{2\pi} \int_{-1}^{1} \frac{\log\left[w_{n,B}^2(s)\sqrt{1 - s^2}\right]}{\sqrt{1 - s^2}}\, ds\right). \tag{12.9}$$

*(b)*

$$\sqrt{\frac{\pi}{2}} p_m \left( w_{n,B}^2, x \right) w_{n,B} (x) \left( 1 - x^2 \right)^{1/4} = \cos \left( m\theta + \Gamma \left( f_n; \theta \right) \right). \tag{12.10}$$

*(c)*

$$\sqrt{\frac{\pi}{2}} p'_m \left( w_{n,B}^2, x \right) w_{n,B} (x) \left( 1 - x^2 \right)^{3/4}$$

$$= \left( m + \Gamma' \left( f_n; \theta \right) \right) \sin \left( m\theta + \Gamma \left( f_n; \theta \right) \right) + \frac{1}{2} \frac{f'_n (\theta)}{f_n (\theta)} \cos \left( m\theta + \Gamma \left( f_n; \theta \right) \right). \tag{12.11}$$

*(d)*

$$\pi \lambda_m^{-1} \left( w_{n,B}^2, x \right) w_{n,B}^2 (x) \left( 1 - x^2 \right)^{1/2}$$

$$= m - \frac{1}{2} + \Gamma' \left( f_n; \theta \right) + \frac{1}{2\sqrt{1-x^2}} \sin \left( (2m-1) \theta + 2\Gamma \left( f_n; \theta \right) \right). \tag{12.12}$$

*Proof.* We apply the results of Appendix B in [25] with a polynomial $S = S_n$ of degree $2q_n$ there and $n$ there replaced by $m \geq q_n$.
(a) This is (a) in [25, p. 435, Theorem B.1].
(b) This is (c) in [25, p. 436, Theorem B.1(c)].
(c) This is (a) of Theorem B.4 in [25, p. 439, Thm. B.4(a)].
(d) This is (b) of Theorem B.4 in [25, p. 440, Thm. B.4(b)]. □
    Recall our notation

$$\psi_n (t) = e^{-nQ_n(t)} w_{n,B}^{-1} (t).$$

Our next lemma provides identities connecting $\Gamma \left( f_n; \cdot \right)$ or $\Gamma \left( F_n; \cdot \right)$ and $\sigma_{Q_n,1}$:

**Lemma 12.2.** *(a) For $x = \cos \theta$,*

$$n + \Gamma' \left( f_n; \theta \right) = n\pi \sqrt{1 - x^2} \sigma_{Q_n} (x)$$

$$+ \frac{1}{\pi} PV \int_{-1}^{1} \frac{\left[ \psi'_n (s) / \psi_n (s) \right] \sqrt{1 - s^2}}{s - x} ds + \frac{1}{2}. \tag{12.13}$$

*(b) Let*

$$F_n (\theta) = e^{-2nQ_n(\cos \theta)} |\sin \theta|. \tag{12.14}$$

*Then*

$$n\theta + \Gamma\left(F_n; \theta\right) = n\pi \int_x^1 \sigma_{Q_n}\left(t\right) dt + \frac{\theta}{2} - \frac{\pi}{4}. \tag{12.15}$$

*Proof.* (a) Now

$$\log \psi_n\left(t\right) = -nQ_n\left(t\right) - \log w_{n,B}\left(t\right),$$

so

$$\frac{d}{dt} \log \psi_n\left(t\right) = -nQ'_n\left(t\right) - \frac{d}{dt} \log w_{n,B}\left(t\right).$$

Then from (12.7),

$$\Gamma'\left(f_n; \theta\right) = -\frac{1}{2\pi} PV \int_{-1}^1 \left\{\frac{d}{ds}\left(\log w_{n,B}^2\left(s\right)\right)\right\} \frac{\sqrt{1-s^2}}{s-x} ds + \frac{1}{2}$$

$$= \frac{n}{\pi} PV \int_{-1}^1 \frac{Q'_n\left(s\right)\sqrt{1-s^2}}{s-x} ds$$

$$+ \frac{1}{\pi} PV \int_{-1}^1 \frac{\left[\psi'_n\left(s\right)/\psi_n\left(s\right)\right]\sqrt{1-s^2}}{s-x} ds + \frac{1}{2}. \tag{12.16}$$

Now from (3.2), with $r = 1$, $a_{\pm n,r} = \pm 1$,

$$\pi\sqrt{1-x^2}\sigma_{Q_n}\left(x\right) = \left(1-x^2\right)\frac{PV}{\pi}\int_{-1}^1 \frac{Q'_n\left(s\right)}{s-x}\frac{ds}{\sqrt{1-s^2}}$$

$$= \frac{PV}{\pi}\int_{-1}^1 \frac{Q'_n\left(s\right)}{s-x}\left(1-s^2+s^2-x^2\right)\frac{ds}{\sqrt{1-s^2}}$$

$$= \frac{PV}{\pi}\int_{-1}^1 \frac{Q'_n\left(s\right)}{s-x}\sqrt{1-s^2}ds + \frac{1}{\pi}\int_{-1}^1 Q'_n\left(s\right)\left(s+x\right)\frac{ds}{\sqrt{1-s^2}}$$

$$= \frac{PV}{\pi}\int_{-1}^1 \frac{Q'_n\left(s\right)}{s-x}\sqrt{1-s^2}ds + 1, \tag{12.17}$$

by our equilibrium relations (3.1). Thus substituting in (12.16) gives

$$\Gamma'\left(f_n; \theta\right) = n\left(\sqrt{1-x^2}\sigma_{Q_n}\left(x\right)\pi - 1\right)$$

$$+ \frac{1}{\pi} PV \int_{-1}^1 \frac{\left[\psi'_n\left(s\right)/\psi_n\left(s\right)\right]\sqrt{1-s^2}}{s-x} ds + \frac{1}{2}.$$

Then (12.13) follows.

(b) Let $x \in (-1, 1)$, $x = \cos\theta$, $\theta \in (0, \pi)$. As at (12.7), we obtain from [25, p. 441, eqn. (23)],

$$
\begin{aligned}
\Gamma'(F_n; \theta) &= -\frac{1}{2\pi} PV \int_{-1}^{1} \left[ \frac{d}{ds} \left( \log\left[ e^{-2nQ_n(s)} \right] \right) \right] \frac{\sqrt{1-s^2}}{s-x} ds + \frac{1}{2} \\
&= \frac{n}{\pi} PV \int_{-1}^{1} Q_n'(s) \frac{\sqrt{1-s^2}}{s-x} ds + \frac{1}{2} \\
&= n \left( \sqrt{1-x^2} \sigma_{Q_n}(x)\, \pi - 1 \right) + \frac{1}{2},
\end{aligned}
\tag{12.18}
$$

by (12.17). Then integrating,

$$
\begin{aligned}
\int_{x}^{1} \left( n + \Gamma'(F_n; \arccos t) \right) \frac{dt}{\sqrt{1-t^2}} &= n\pi \int_{x}^{1} \sigma_{Q_n}(t)\, dt + \frac{1}{2} \int_{x}^{1} \frac{dt}{\sqrt{1-t^2}} \\
\Rightarrow \int_{0}^{\theta} \left( n + \Gamma'(F_n; s) \right) ds &= n\pi \int_{x}^{1} \sigma_{Q_n}(t)\, dt + \frac{\theta}{2} \\
\Rightarrow n\theta + \Gamma(F_n; \theta) - \Gamma(F_n; 0) &= n\pi \int_{x}^{1} \sigma_{Q_n}(t)\, dt + \frac{\theta}{2}.
\end{aligned}
\tag{12.19}
$$

We turn to evaluating $\Gamma(F_n; 0)$. Now by [25, p. 433, eqn. (41)]

$$
\begin{aligned}
\Gamma(F_n; \theta) &= \Gamma\left( e^{-2n\breve{Q}_n}; \theta \right) + \frac{1}{2} \Gamma\left( \sin^2 \cdot\, ; \theta \right) \\
&= \Gamma\left( e^{-2n\breve{Q}_n}; \theta \right) + \frac{\theta}{2} - \frac{\pi}{4}.
\end{aligned}
\tag{12.20}
$$

Here, by Lemma B.5(a) in [25, p. 441, eqn. (21)], with $x = \cos\theta$,

$$
\Gamma\left( e^{-2n\breve{Q}_n}; \theta \right) = -\frac{n\sqrt{1-x^2}}{\pi} \int_{-1}^{1} \frac{Q_n'(s) - Q_n'(x)}{s-x} \frac{ds}{\sqrt{1-s^2}}
$$

and since $Q_n'$ satisfies a Lipschitz condition of order $\alpha_1 > \frac{1}{2}$ near 1, we have

$$
\Gamma\left( e^{-2n\breve{Q}_n}; 0 \right) = \lim_{x \to 1-} \Gamma\left( e^{-2n\breve{Q}_n}; \theta \right) = 0.
$$

Then (12.20) gives

$$
\Gamma(F_n; 0) = -\frac{\pi}{4}.
\tag{12.21}
$$

Substituting into (12.19) gives (12.15).                                                 □

We also need an estimate on $p_m \left( w_{n,B}^2, \cdot \right)$ in the complex plane:

**Lemma 12.3.** *Let* $m \geq q_n$ *and*

$$\phi(z) = z + \sqrt{z^2 - 1}. \tag{12.22}$$

*Then for* $z \in \mathbb{C} \setminus [-1, 1]$,

$$\left| p_m \left( w_{n,B}^2, z \right) \Big/ \left\{ \frac{1}{\sqrt{\pi}} \phi(z)^m D^{-2} \left( \breve{w}_{n,B}; \phi(z)^{-1} \right) \left( 1 - \phi(z)^{-2} \right)^{-1/2} \right\} - 1 \right|$$

$$\leq |\phi(z)|^{2q_n - 2m - 2}.$$

*Proof.* This is the special case $p = 2$ of Theorem A.1(d) in [25, p. 422]. Note that there $V_p = V_2 = w_{n,B}$ and $\kappa_2 = \sqrt{\frac{\pi}{2}}$.                                     □

# Chapter 13
# Asymptotics of Orthonormal Polynomials

In this chapter, we establish asymptotics of orthogonal polynomials on and off the interval of orthogonality, as well as their first derivative, their leading coefficients, their recurrence coefficients, their zeros, and the associated Christoffel functions. We first state the theorems, and then prove them sequentially. Throughout, we assume that $\{Q_n\} \in \mathcal{Q}$, as in Definition 1.1.

**Theorem 13.1.** *Let $\tau_1 \in \left(0, \frac{\alpha}{50}\right)$. For $|n - m| \le \frac{1}{2} n^{1/3 - 2\tau_1}$, and some $C > 0$, the leading coefficient $\gamma_{n,m}$ of $p_{n,m}$ satisfies*

$$\gamma_{n,m} = \frac{2^m}{\sqrt{\pi}} \exp\left(\frac{n}{\pi} \int_{-1}^{1} Q_n(x) \frac{dx}{\sqrt{1 - x^2}}\right) \left(1 + O\left(n^{-C}\right)\right). \tag{13.1}$$

*Moreover, if $w_{n,B}$ is as in Theorem 11.1, then for some $C > 0$,*

$$\gamma_{n,m} = \gamma_m\left(w_{n,B}^2\right)\left(1 + O\left(n^{-C}\right)\right). \tag{13.2}$$

The next result focuses on asymptotics inside and outside $[-1, 1]$:

**Theorem 13.2.** *Let $\varepsilon \in \left(0, \frac{1}{3}\right)$. Let $\{F_n\}$ be defined by (12.14) and $\{\Gamma(F_n; \cdot)\}$ be defined by (12.15). For $n \ge 1$, let*

$$|m - n| \le n^{1/3 - \varepsilon}. \tag{13.3}$$

*There exists $\tau_1 \in \left(0, \frac{\varepsilon}{4}\right)$ with the following properties:*
*(a) For $|x| \le 1 - n^{-\tau_1}$, and $\theta = \arccos x$,*

$$\sqrt{\frac{\pi}{2}} p_{n,m}(x) e^{-nQ_n(x)} \left(1 - x^2\right)^{1/4}$$

$$= \cos\left(m\theta + \Gamma(F_n; \theta)\right) + O\left(n^{-\tau_1}\right)$$

© The Author(s) 2018
E. Levin, D.S. Lubinsky, *Bounds and Asymptotics for Orthogonal Polynomials for Varying Weights*, SpringerBriefs in Mathematics,
https://doi.org/10.1007/978-3-319-72947-3_13

$$= \cos \left( (m - n) \, \theta + n\pi \int_x^1 \sigma_{Q_n} (t) \, dt + \frac{\theta}{2} - \frac{\pi}{4} \right)$$

$$+ O \left( n^{-\tau_1} \right). \tag{13.4}$$

(b) For $|x| \le 1 - n^{-\tau_1}$,

$$\frac{1}{n} \sqrt{\frac{\pi}{2}} p'_{n,m} (x) \, e^{-nQ_n(x)} \left( 1 - x^2 \right)^{1/4}$$

$$= \pi \sigma_{Q_n} (x) \sin \left( m\theta + \Gamma \left( F_n; \theta \right) \right)$$

$$+ Q'_n (\cos \theta) \cos \left( m\theta + \Gamma \left( F_n; \theta \right) \right) + O \left( n^{-\tau_1/3} \right)$$

$$= \pi \sigma_{Q_n} (x) \sin \left( (m - n) \, \theta + n\pi \int_x^1 \sigma_{Q_n} (t) \, dt + \frac{\theta}{2} - \frac{\pi}{4} \right)$$

$$+ Q'_n (x) \cos \left( (m - n) \, \theta + n\pi \int_x^1 \sigma_{Q_n} (t) \, dt + \frac{\theta}{2} - \frac{\pi}{4} \right) + O \left( n^{-\tau_1/3} \right). \tag{13.5}$$

(c) For $\mathrm{dist} \, (z, [-1, 1]) \ge n^{-\tau_1/2}$,

$$\left| p_{n,m} (z) \, / \left\{ \frac{1}{\sqrt{2\pi}} \phi (z)^m D^{-1} \left( F_n; \phi (z)^{-1} \right) \right\} - 1 \right| \le C n^{-\tau_1/2}. \tag{13.6}$$

Next, we turn to Christoffel functions:

**Theorem 13.3.** *There exists $\tau_1 > 0$ such that for $|x| \le 1 - n^{-\tau_1}$,*

$$\lambda_n^{-1} \left( e^{-2nQ_n}, x \right) e^{-2nQ_n(x)} = n\sigma_{Q_n} (x) + O \left( n^{1-\tau_1} \right). \tag{13.7}$$

Recall that the three term recurrence relation for $p_{n,m}$ has the form

$$x p_{n,m} (x) = A_{n,m} p_{n,m+1} (x) + B_{n,m} p_{n,m} (x) + A_{n,m-1} p_{n,m-1} (x).$$

We prove:

**Theorem 13.4.** *Let $\varepsilon \in \left( 0, \frac{1}{3} \right)$. Uniformly for $m = m (n)$ satisfying (13.3), and some $\tau_1 > 0$,*

$$A_{n,m} = \frac{1}{2} + O \left( n^{-\tau_1} \right) \text{ and } B_{n,m} = O \left( n^{-\tau_1} \right). \tag{13.8}$$

Our final result in this chapter concerns asymptotics of zeros of orthogonal polynomials. For $n \ge 1$, define $g_n : [-1, 1] \to \mathbb{R}$ by

$$g_n (x) = \int_x^1 \sigma_{Q_n} (t) \, dt + \frac{\arccos x}{2\pi n} + \frac{1}{4n}, \quad x \in [-1, 1]. \tag{13.9}$$

Then

$$g_n(-1) = 1 + \frac{3}{4n} \text{ and } g_n(1) = \frac{1}{4n},$$

and moreover,

$$g_n'(x) = -\sigma_{Q_n}(x) - \frac{1}{2\pi n \sqrt{1-x^2}} < 0, \quad x \in (-1,1). \tag{13.10}$$

It follows that there is a unique root $y_{jn} \in (-1,1)$ of

$$g_n(y_{jn}) = \frac{j}{n}, \quad 1 \le j \le n. \tag{13.11}$$

Moreover, since $\sigma_{Q_n}(x) \sim \sqrt{1-x^2}$ and $|g_n'| \sim \sigma_{Q_n}$ uniformly in such $n,x$,

$$y_{jn} - y_{j+1,n} \sim \frac{1}{n\sqrt{1-y_{jn}^2}}, \tag{13.12}$$

uniformly for $j$ with $|y_{jn}| \le 1 - \frac{C}{n}$, some large enough $C$.

**Theorem 13.5.** *There exists $\tau_2 > 0$ and $n_0$ such that for $n \ge n_0$,*
  *(a) Uniformly for $k$ with $|x_{kn}| \le 1 - n^{-\tau_2}$, there is a $j = j(k,n)$ such that*

$$n(x_{kn} - y_{jn}) = O(n^{-\tau_2}). \tag{13.13}$$

*Moreover for each $j$, there is exactly one $k$ for which this last estimate holds.*
  *(b) Uniformly for $k$ with $|x_{kn}| \le 1 - n^{-\tau_2}$,*

$$n\sigma_{Q_n}(x_{kn})(x_{kn} - x_{k+1,n}) = 1 + O(n^{-\tau_2}). \tag{13.14}$$

We begin the proofs of the theorems:

**Lemma 13.6.** *Let $r^*$ be as in Definition 1.1, and*

$$H_n(\lambda) = n\left\{ \frac{1}{\pi} \int_{-1}^{1} \frac{Q_n(\lambda x)}{\sqrt{1-x^2}} dx - \log \lambda \right\}, \quad 0 \le \lambda \le r^*. \tag{13.15}$$

*Then*

$$H_n(\lambda) \ge H_n(1) = \frac{n}{\pi} \int_{-1}^{1} \frac{Q_n(x)}{\sqrt{1-x^2}} dx. \tag{13.16}$$

*Proof.* This follows from the maximum property of the $\mathcal{F}$-functional [44, p. 194, Thm. IV 1 5]: for $0 \le \lambda \le r^*$,

$$\log \frac{\lambda}{2} - \int_{-\lambda}^{\lambda} \frac{Q_n(t)}{\pi\sqrt{\lambda^2 - t^2}} dt \le \log \frac{1}{2} - \int_{-1}^{1} \frac{Q_n(t)}{\pi\sqrt{1-t^2}} dt,$$

since $[-1, 1]$ is the support of the equilibrium measure for $Q_n$. This last inequality is easily reformulated as (13.16). $\qquad\square$

*Proof of Theorem* 13.1. We first prove the bound:

$$\gamma_{n,m}^{-2} \leq \left(1 + O\left(n^{-C}\right)\right) \left(\frac{2^m}{\sqrt{2\pi}}\right)^{-2}$$

$$\times \exp\left(\frac{1}{\pi} \int_{-1}^{1} \left\{-2nQ_n\left(x\right) + \log\sqrt{1 - x^2}\right\} \frac{dx}{\sqrt{1 - x^2}}\right). \qquad (13.17)$$

To prove this, for $n \geq 1$, let

$$\rho_n = 1 + \left(\frac{A\log n}{n}\right)^{2/3},$$

where $A$ is large enough. Using the extremal property of leading coefficients, and our restricted range inequality Theorem 4.2(c) with $T = 1$, $S = n^{1/3}$, and $R = (A\log n)^{2/3}$ with large enough $A > 0$, we have for some $C > 0$,

$$\gamma_{n,m}^{-2} = \inf_{\deg(P)\leq m-1} \int_{I_n} \left(x^m - P\left(x\right)\right)^2 e^{-2nQ_n(x)} dx$$

$$\leq \left(1 + O\left(n^{-C}\right)\right) \inf_{\deg(P)\leq m-1} \int_{-\rho_n}^{\rho_n} \left(x^m - P\left(x\right)\right)^2 e^{-2nQ_n(x)} dx$$

$$= \left(1 + O\left(n^{-C}\right)\right) \rho_n^{2m+1} \inf_{\deg(P)\leq m-1} \int_{-1}^{1} \left(t^m - P\left(t\right)\right)^2 e^{-2nQ_n(\rho_n t)} dt.$$

Next, Theorem 11.2 gives polynomials $S_n$ of degree $\leq 2n - n^{1/3 - 2\tau_1}$ and Bernstein-Szegő weights $w_{n,B}$, with

$$\hat{\psi}_n\left(t\right) = e^{-nQ_n(\rho_n t)} w_{n,B}^{-1}\left(t\right) \leq 1, \quad t \in [-1, 1].$$

Then using this inequality and (12.9),

$$\gamma_{n,m}^{-2} \leq \left(1 + O\left(n^{-C}\right)\right) \rho_n^{2m+1} \inf_{\deg(P)\leq m-1} \int_{-1}^{1} \left(t^m - P\left(t\right)\right)^2 w_{n,B}^2\left(t\right) dt$$

$$= \left(1 + O\left(n^{-C}\right)\right) \rho_n^{2m+1} \gamma_m^{-2}\left(w_{n,B}^2\right)$$

$$= \left(1 + O\left(n^{-C}\right)\right) \rho_n^{2m+1} \left(\frac{2^m}{\sqrt{2\pi}}\right)^{-2}$$

$$\times \exp\left(\frac{1}{\pi} \int_{-1}^{1} \log\left[w_{n,B}^2\left(x\right) \sqrt{1 - x^2}\right] \frac{dx}{\sqrt{1 - x^2}}\right)$$

$$= \left(1 + O\left(n^{-C}\right)\right) \rho_n^{2m+1} \left(\frac{2^m}{\sqrt{2\pi}}\right)^{-2}$$

$$\times \exp\left(\frac{1}{\pi} \int_{-1}^{1} \left\{-2nQ_n\left(\rho_n x\right) - \log\left[\hat{\psi}_n\left(x\right)^2\right] + \log\sqrt{1-x^2}\right\} \frac{dx}{\sqrt{1-x^2}}\right)$$

$$= \left(1 + O\left(n^{-C}\right)\right) \rho_n^{2m+1} \left(\frac{2^m}{\sqrt{2\pi}}\right)^{-2}$$

$$\times \exp\left(\frac{1}{\pi} \int_{-1}^{1} \left\{-2nQ_n\left(\rho_n x\right) + \log\sqrt{1-x^2}\right\} \frac{dx}{\sqrt{1-x^2}}\right), \tag{13.18}$$

by (11.16) of Theorem 11.2. Now we apply Lemma 13.6, with $H_n$ as in (13.15),

$$\rho_n^{2m} \exp\left(-\frac{1}{\pi} \int_{-1}^{1} 2nQ_n\left(\rho_n x\right) \frac{dx}{\sqrt{1-x^2}}\right)$$

$$= \rho_n^{2(m-n)} \exp\left(-2H_n\left(\rho_n\right)\right)$$

$$\leq \exp\left(O\left((m-n)\left(\frac{\log n}{n}\right)^{2/3}\right)\right) \exp\left(-2H_n\left(1\right)\right)$$

$$= \exp\left(-\frac{1}{\pi} \int_{-1}^{1} 2nQ_n\left(x\right) \frac{dx}{\sqrt{1-x^2}}\right) \left(1 + O\left(n^{-C}\right)\right),$$

some $C > 0$, recall that $|m - n| = O\left(n^{1/3}\right)$. Substituting this in (13.18) gives (13.17). Next, we prove the matching lower bound

$$\gamma_{n,m}^{-2} \geq \left(1 + O\left(n^{-C}\right)\right) \left(\frac{2^m}{\sqrt{2\pi}}\right)^{-2}$$

$$\times \exp\left(\frac{1}{\pi} \int_{-1}^{1} \left\{-2nQ_n\left(x\right) + \log\sqrt{1-x^2}\right\} \frac{dx}{\sqrt{1-x^2}}\right). \tag{13.19}$$

Here we choose $w_{n,B}$ as in Theorem 11.3. By that theorem, followed by (12.9),

$$\gamma_{n,m}^{-2} \geq \inf_{\deg(P) \leq m-1} \int_{-1}^{1} \left(x^m - P\left(x\right)\right)^2 e^{-2nQ_n(x)} dx$$

$$\geq \inf_{\deg(P) \leq m-1} \int_{-1}^{1} \left(t^m - P\left(t\right)\right)^2 w_{n,B}^2\left(t\right) dt$$

$$= \gamma_m \left(w_{n,B}^2\right)^{-2}$$

$$= \left(\frac{2^m}{\sqrt{2\pi}}\right)^{-2} \exp\left(\frac{1}{\pi} \int_{-1}^{1} \log\left[w_{n,B}^2\left(x\right)\sqrt{1-x^2}\right] \frac{dx}{\sqrt{1-x^2}}\right)$$

$$= \left(\frac{2^m}{\sqrt{2\pi}}\right)^{-2} \exp\left(\frac{1}{\pi}\int_{-1}^{1}\left\{-2nQ_n(x) - \log\left[\psi_n(x)^2\right]\right.\right.$$

$$\left.\left. + \log\sqrt{1-x^2}\right\}\right)\frac{dx}{\sqrt{1-x^2}}$$

$$= \left(1 + O\left(n^{-C}\right)\right)\left(\frac{2^m}{\sqrt{2\pi}}\right)^{-2}\exp\left(\frac{1}{\pi}\int_{-1}^{1}\left\{-2nQ_n(x)\right.\right.$$

$$\left.\left. + \log\sqrt{1-x^2}\right\}\frac{dx}{\sqrt{1-x^2}}\right),$$

by (11.18). So we have (13.19).

Finally, (13.1) follows from (13.17) and (13.19), and the classic potential theory identity

$$\frac{1}{\pi}\int_{-1}^{1}\frac{\log(1\pm x)}{\sqrt{1-x^2}}dx = -\log 2.$$

Also (13.2) follows easily using Theorem 11.1.                                    □

Next, we turn to the proof of Theorem 13.2. That theorem refers to *some* $\tau_1 \in \left(0, \frac{\varepsilon}{4}\right)$. We shall sometimes decrease the parameter $\tau_1$. This has the effect of decreasing the range $|x| \le 1 - n^{-\tau_1}$ and increasing the error estimate $O\left(n^{-\tau_1}\right)$. Our restriction that $\tau_1 < \frac{\varepsilon}{4}$ ensures that (13.3) is compatible with the requirement $m \ge q_n = \frac{1}{2}\deg(S_n)$ in Lemma 12.1 and our use of Theorem 11.1. There $S_n$ has degree $\le 2n - n^{1/3 - 2\tau_1}$, which is compatible with (13.3) if

$$m \ge n - n^{1/3 - \varepsilon} \ge n - \frac{1}{2}n^{1/3 - 2\tau_1}$$

$$\Leftrightarrow n^{\varepsilon} \ge 2n^{2\tau_1}.$$

Our hypothesis $\tau_1 < \frac{\varepsilon}{4}$ ensures that this last inequality holds for large enough $n$.

We now show that $p_{n,m}(x)$ is close to $p_m\left(w_{n,B}^2, x\right)$ when $w_{n,B}^2$ is as in Theorem 11.1:

**Lemma 13.7.** *Assume* (13.3). *Let* $\{w_{n,B}\}$ *be as in Theorem 11.1, and*

$$\pi_{n,m}(x) = p_{n,m}(x) - \frac{\gamma_{n,m}}{\gamma_m\left(w_{n,B}^2\right)}p_m\left(w_{n,B}^2, x\right). \qquad (13.20)$$

*Then for some* $\tau_1 > 0$,
*(a)*

$$\sup_{|x|\le 1 - n^{-\tau_1}} |\pi_{n,m}(x)| w_{n,B}(x)\left(1 - x^2\right)^{1/4} \le Cn^{-\tau_1/3}. \qquad (13.21)$$

(b)

$$\sup_{1-n^{-\tau_1} \le |x| \le 1} |\pi_{n,m}(x)| \, w_{n,B}(x) \, (1-x^2)^{3/2} \le Cn^{5\tau_1}. \tag{13.22}$$

(c)

$$\sup_{|x| \le 1-n^{-\tau_1/48}} |\pi'_{n,m}(x)| \, e^{-nQ_n(x)} \le Cn^{1-\tau_1/6}. \tag{13.23}$$

(d)

$$\sup_{|x| \le 1-n^{-\tau_1/48}} \frac{1}{n} \left| \lambda_n^{-1}\left(e^{-2nQ_n}, x\right) - \lambda_n^{-1}\left(w_{n,B}^2, x\right) \right| e^{-2nQ_n(x)} \le Cn^{-\tau_1}. \tag{13.24}$$

*Proof.* (a) We use Korous' method: by orthogonality,

$$\pi_{n,m}(x) = \int_{-1}^{1} \pi_{n,m}(t) \, K_m\left(w_{n,B}^2, x, t\right) w_{n,B}^2(t) \, dt$$

$$= \int_{-1}^{1} p_{n,m}(t) \, K_m\left(w_{n,B}^2, x, t\right) w_{n,B}^2(t) \, dt$$

$$= \int_{-1}^{1} p_{n,m}(t) \, K_m\left(w_{n,B}^2, x, t\right)$$

$$\times \left[ w_{n,B}^2(t) - w_{n,B}^2(x) \, e^{2n(Q_n(x)-Q_n(t))} \right] dt.$$

Recall that

$$\psi_n(t) = w_{n,B}^{-1}(t) \, e^{-nQ_n(t)},$$

and let

$$\Delta_n(x, t) = \frac{\left[1 - \left(\frac{\psi_n(t)}{\psi_n(x)}\right)^2\right]}{x - t}.$$

Then using the Christoffel-Darboux formula,

$$\pi_{n,m}(x) = \int_{-1}^{1} p_{n,m}(t) \, K_m\left(w_{n,B}^2, x, t\right) w_{n,B}^2(t) \left[1 - \left(\frac{\psi_n(t)}{\psi_n(x)}\right)^2\right] dt$$

$$= \frac{\gamma_{m-1}\left(w_{n,B}^2\right)}{\gamma_m\left(w_{n,B}^2\right)}$$

$$\times \left\{ \begin{array}{l} p_m\left(w_{n,B}^2, x\right) \int_{-1}^{1} p_{n,m}(t) \, p_{m-1}\left(w_{n,B}^2, t\right) w_{n,B}^2(t) \, \Delta_n(x, t) \, dt \\ -p_{m-1}\left(w_{n,B}^2, x\right) \int_{-1}^{1} p_{n,m}(t) \, p_m\left(w_{n,B}^2, t\right) w_{n,B}^2(t) \, \Delta_n(x, t) \, dt \end{array} \right\}.$$

Here by (12.9), $\frac{\gamma_{m-1}(w_{n,B}^2)}{\gamma_m(w_{n,B}^2)} = \frac{1}{2}$. Next from (12.10), for $k = m-1, m$, and $x \in (-1, 1)$,

$$\left| p_k \left( w_{n,B}^2, x \right) \right| w_{n,B} (x) \left( 1 - x^2 \right)^{1/4} \leq \sqrt{\frac{2}{\pi}}. \tag{13.25}$$

Also, by Theorem 7.1,

$$\sup_{t \in [-1,1]} \left| p_{n,m} (t) \right| e^{-nQ_n(t)} \left( 1 - t^2 \right)^{1/4} \leq C. \tag{13.26}$$

Then for $|x| < 1$,

$$\left| \pi_{n,m} (x) \right| w_{n,B} (x) \left( 1 - x^2 \right)^{1/4}$$

$$\leq C \int_{-1}^{1} e^{nQ_n(t)} w_{n,B} (t) \left| \Delta_n (x, t) \right| \frac{dt}{\sqrt{1 - t^2}}$$

$$= C \int_{-1}^{1} \frac{\left| 1 - \left( \frac{\psi_n(t)}{\psi_n(x)} \right)^2 \right| \psi_n (t)^{-1}}{|x - t|} \frac{dt}{\sqrt{1 - t^2}}.$$

Now Theorem 11.1(f) gives the result.

(b) Here we apply Theorem 11.1(h) in the above inequality: for $1 - n^{-\tau_1} \leq |x| \leq 1$,

$$\int_{-1}^{1} \frac{\left| 1 - \left( \frac{\psi_n(t)}{\psi_n(x)} \right)^2 \right| \psi_n (t)^{-1}}{|x - t|} \frac{dt}{\sqrt{1 - t^2}} \leq C n^{5\tau_1} \left( 1 - |x| \right)^{-5/4}. \tag{13.27}$$

(c) Observe first that for $x \in [-1, 1]$,

$$\left| \pi_{n,m} (x) \right| e^{-nQ_n(x)} = \left| \pi_{n,m} (x) \right| w_{n,B} (x) \psi_n (x) \leq \left| \pi_{n,m} (x) \right| w_{n,B} (x), \tag{13.28}$$

as follows from (11.4). We see from (b) that for $1 - n^{-\tau_1} \leq |x| \leq 1$,

$$\left( 1 - x^2 \right)^8 \left| \pi_{n,m} (x) \right| w_{n,B} (x) \leq C n^{-\tau_1}.$$

Also for $|x| \leq 1 - n^{-\tau_1}$, (a) gives

$$\left( 1 - x^2 \right)^8 \left| \pi_{n,m} (x) \right| w_{n,B} (x) \leq C n^{-\tau_1/3}.$$

These last two inequalities, the fact that $e^{-nQ_n} \leq w_{n,B}$ in $[-1, 1]$, and our restricted range inequality (4.4) give

$$\left( 1 - x^2 \right)^8 \left| \pi_{n,m} (x) \right| e^{-nQ_n(x)} \leq C n^{-\tau_1/3}, \quad x \in I_n.$$

By our Markov-Bernstein inequality Theorem 8.1(b), for $|x| < 1$,

$$\left| \frac{d}{dx} \left\{ (1 - x^2)^8 \, \pi_{n,m}(x) \right\} \right| e^{-nQ_n(x)} \le Cn^{1-\tau_1/3}.$$

Then

$$(1 - x^2)^8 \, \left| \pi'_{n,m}(x) \right| e^{-nQ_n(x)} \le Cn^{1-\tau_1/3} + 16 \, (1 - x^2)^7 \, |\pi_{n,m}(x)| \, e^{-nQ_n(x)}$$

$$\Rightarrow \left| \pi'_{n,m}(x) \right| e^{-nQ_n(x)} \le \frac{Cn^{1-\tau_1/3}}{(1 - x^2)^8} + \frac{16}{1 - x^2} |\pi_{n,m}(x)| \, e^{-nQ_n(x)} \le Cn^{1-\tau_1/6}$$

for $|x| \le 1 - n^{-\tau_1/48}$.

(d) This follows in a fairly straightforward but technical fashion from our earlier asymptotics. We provide some details. Recall the formulae

$$\lambda_n^{-1} \left( e^{-2nQ_n}, x \right) = \frac{\gamma_{n,n-1}}{\gamma_{n,n}} \left( p'_{n,n}(x) \, p_{n,n-1}(x) - p_{n,n}(x) \, p'_{n,n-1}(x) \right);$$

$$\lambda_n^{-1} \left( w_{n,B}^2, x \right) = \frac{\gamma_{n-1} \left( w_{n,B}^2 \right)}{\gamma_n \left( w_{n,B}^2 \right)} \left( p'_n \left( w_{n,B}^2, x \right) p_{n-1} \left( w_{n,B}^2, x \right) \right.$$

$$\left. - p_n \left( w_{n,B}^2, x \right) p'_{n-1} \left( w_{n,B}^2, x \right) \right). \tag{13.29}$$

We also use that by (12.9), (13.1), and (13.2), for $m = n - 1, n$,

$$\frac{\gamma_{n-1} \left( w_{n,B}^2 \right)}{\gamma_n \left( w_{n,B}^2 \right)} = \frac{1}{2} \quad \text{and} \quad \frac{\gamma_{n,n-1}}{\gamma_{n,n}} = \frac{1}{2} + O \left( n^{-C} \right) \quad \text{and}$$

$$\frac{\gamma_{n,m}}{\gamma_m \left( w_{n,B}^2 \right)} = 1 + O \left( n^{-C} \right). \tag{13.30}$$

Let

$$\pi_{n,m}^{\#}(x) = p_{n,m}(x) - p_m \left( w_{n,B}^2, x \right)$$

$$= \pi_{n,m}(x) + O \left( n^{-C} \right) p_m \left( w_{n,B}^2, x \right).$$

Recall the bounds (13.21), (13.25) and that $e^{-nQ_n} \le w_{n,B}$. Then for some small enough $\tau_0$ and $m = n - 1, n$,

$$\sup_{|x| \le 1 - n^{-\tau_1}} \left| \pi_{n,m}^{\#}(x) \right| e^{-nQ_n(x)} \le Cn^{-\tau_0}, \tag{13.31}$$

and much as in (c),

$$\sup_{|x| \le 1 - n^{-\tau_1/48}} \left| \pi_{n,m}^{\#\prime}(x) \right| e^{-nQ_n(x)} \le Cn^{1-\tau_0}. \tag{13.32}$$

Note too that the bound

$$\sup_{x \in I_n} \left| p_{n,m}(x)(1-x^2)e^{-nQ_n(x)} \right| \le C$$

(which follows from Theorem 7.1 and Theorem 4.2(a)) and Theorem 8.1(b) imply

$$\left| p'_{n,m}(x) \right| e^{-nQ_n(x)} \le C\frac{n}{1-x^2}, \qquad |x| < 1. \tag{13.33}$$

Next some straightforward manipulations give

$$\begin{aligned}
\Psi_n(x) &:= \left( p'_{n,n}(x)\,p_{n,n-1}(x) - p_{n,n}(x)\,p'_{n,n-1}(x) \right) \\
&\quad - \left( p'_n\left(w_{n,B}^2, x\right)p_{n-1}\left(w_{n,B}^2, x\right) - p_n\left(w_{n,B}^2, x\right)p'_{n-1}\left(w_{n,B}^2, x\right) \right) \\
&= p'_{n,n}(x)\,\pi^{\#}_{n,n-1}(x) - \pi^{\#}_{n,n}(x)\,p'_{n,n-1}(x) \\
&\quad + \pi^{\#\prime}_{n,n}(x)\,p_{n-1}\left(w_{n,B}^2, x\right) - \pi^{\#\prime}_{n,n-1}(x)\,p_n\left(w_{n,B}^2, x\right).
\end{aligned}$$

Then our estimates (13.25–26) and (13.31–(13.33)) give for $|x| \le 1 - n^{-\tau_1/48}$, and some small enough $\tau_0$,

$$\left| \Psi_n(x) \right| e^{-2nQ_n(x)} \le C\frac{n^{1-\tau_0}}{1-x^2}.$$

Combining this and (13.29–13.30) gives the result.                                    $\square$

*Proof of Theorem* 13.2(a). We use the Bernstein-Szegő weights of Theorem 11.1. Now from (11.2), (12.2), and (12.14),

$$f_n(\theta) = F_n(\theta)\,\psi_n^{-2}(\cos\theta).$$

Using

$$\cos t - \cos\theta = 2\sin\left(\frac{\theta-t}{2}\right)\sin\left(\frac{\theta+t}{2}\right)$$

and recalling (12.5),

$$\begin{aligned}
\Gamma(f_n; \theta) - \Gamma(F_n; \theta) &= \frac{1}{4\pi}\int_{-\pi}^{\pi}\left[\log\psi_n^{-2}(\cos t) - \log\psi_n^{-2}(\cos\theta)\right]\cot\left(\frac{\theta-t}{2}\right)dt \\
&= \frac{1}{2\pi}\int_0^{2\pi}\frac{\log\psi_n^{-2}(\cos t) - \log\psi_n^{-2}(\cos\theta)}{\cos t - \cos\theta}\sin\theta\,dt
\end{aligned}$$

and hence, for $|x| \le 1 - n^{-\tau_1}$ with $x = \cos\theta$,

$$\begin{aligned}
\left| \Gamma(f_n; \theta) - \Gamma(F_n; \theta) \right| &\le C\int_{-1}^{1}\left| \frac{\log\psi_n^{-2}(s) - \log\psi_n^{-2}(x)}{s-x} \right|\frac{ds}{\sqrt{1-s^2}} \\
&< Cn^{-\tau_1/3}, \tag{13.34}
\end{aligned}$$

by (11.9). We substitute this in (12.10):

$$\sqrt{\frac{\pi}{2}} P_m \left( w_{n,B}^2, x \right) w_{n,B}(x) \left( 1 - x^2 \right)^{1/4} = \cos \left( m\theta + \Gamma(f_n; \theta) \right)$$

$$= \cos \left( m\theta + \Gamma(F_n; \theta) \right) + O \left( n^{-\tau_1/3} \right).$$

Then from Lemma 13.7(a), (13.30), and (11.3),

$$p_{n,m}(x) e^{-nQ_n(x)} \left( 1 - x^2 \right)^{1/4} = \sqrt{\frac{2}{\pi}} \cos \left( m\theta + \Gamma(F_n; \theta) \right) + O \left( n^{-\tau_1/3} \right).$$

This gives the first asymptotic in (13.4), except that $\tau_1$ is replaced by $\tau_1/3$. We can just reduce $\tau_1$ to $\tau_1/3$. The second follows from (12.15).  $\square$

*Proof of Theorem 13.2(b).* Recall from Lemma 12.1(c) that

$$\sqrt{\frac{\pi}{2}} P_m' \left( w_{n,B}^2, x \right) w_{n,B}(x) \left( 1 - x^2 \right)^{3/4}$$

$$= \left( m + \Gamma'(f_n; \theta) \right) \sin \left( m\theta + \Gamma(f_n; \theta) \right) + \frac{1}{2} \frac{f_n'(\theta)}{f_n(\theta)} \cos \left( m\theta + \Gamma(f_n; \theta) \right).$$

$$(13.35)$$

Here if $\theta \in (0, \pi)$ is such that $x = \cos \theta$ satisfies $|x| \leq 1 - n^{-\tau_1}$,

$$\frac{f_n'(\theta)}{f_n(\theta)} = \frac{d}{d\theta} \log \left\{ w_{n,B}^2(\cos\theta) |\sin\theta| \right\}$$

$$= -2 \frac{d}{d\theta} \left( \log \psi_n(\cos\theta) + nQ_n(\cos\theta) \right) + \cot\theta$$

$$= 2 \frac{\psi_n'(\cos\theta)}{\psi_n(\cos\theta)} \sin\theta + 2nQ_n'(\cos\theta) \sin\theta + \cot\theta$$

$$= 2nQ_n'(\cos\theta) \sin\theta + O \left( n^{1-\tau_1} \right),$$

by (11.3) and (11.5). Also, by (12.13),

$$n + \Gamma'(f_n; \theta)$$

$$= n\pi \sqrt{1 - x^2} \sigma_{Q_n}(x) + \frac{1}{\pi} PV \int_{-1}^1 \frac{\frac{\psi_n'(s)}{\psi_n(s)}(1 - s^2) - \frac{\psi_n'(x)}{\psi_n(x)}(1 - x^2)}{s - x} \frac{ds}{\sqrt{1 - s^2}}$$

$$= n\pi \sqrt{1 - x^2} \sigma_{Q_n}(x) + O \left( n^{1-\tau_1/3} \right),$$

by (11.11), so, recalling (13.3),

$$m + \Gamma'(f_n; \theta) = n\pi \sqrt{1 - x^2} \sigma_{Q_n}(x) + O \left( n^{1-\tau_1/3} \right) + O \left( n^{1/3} \right)$$

$$= n\pi \sqrt{1 - x^2} \sigma_{Q_n}(x) + O \left( n^{1-\tau_1/3} \right).$$

Then from (13.34) and (13.35), and (1.17),

$$\frac{1}{n}\sqrt{\frac{\pi}{2}}p'_m\left(w^2_{n,B},x\right)w_{n,B}\left(x\right)\left(1-x^2\right)^{3/4}$$

$$= \left(\pi\sqrt{1-x^2}\sigma_{Q_n}\left(x\right)+O\left(n^{-\tau_1/3}\right)\right)\sin\left(m\theta+\Gamma\left(F_n;\theta\right)+O\left(n^{-\tau_1/3}\right)\right)$$

$$+ \left(Q'_n\left(\cos\theta\right)\sin\theta+O\left(n^{-\tau_1}\right)\right)\cos\left(m\theta+\Gamma\left(F_n;\theta\right)+O\left(n^{-\tau_1/3}\right)\right)$$

$$= \pi\sqrt{1-x^2}\sigma_{Q_n}\left(x\right)\sin\left(m\theta+\Gamma\left(F_n;\theta\right)\right)$$

$$+ Q'_n\left(\cos\theta\right)\sin\theta\cos\left(m\theta+\Gamma\left(F_n;\theta\right)\right)+O\left(n^{-\tau_1/3}\right)$$

and hence also

$$\frac{1}{n}\sqrt{\frac{\pi}{2}}p'_m\left(w^2_{n,B},x\right)w_{n,B}\left(x\right)\left(1-x^2\right)^{1/4}$$

$$= \pi\sigma_{Q_n}\left(x\right)\sin\left(m\theta+\Gamma\left(F_n;\theta\right)\right)+Q'_n\left(\cos\theta\right)\cos\left(m\theta+\Gamma\left(F_n;\theta\right)\right)$$

$$+ O\left(\frac{n^{-\tau_1/3}}{\sqrt{1-x^2}}\right).$$

In view of Lemma 13.7(c) and (13.30), also

$$\frac{1}{n}\sqrt{\frac{\pi}{2}}p'_{n,m}\left(x\right)e^{-nQ_n(x)}\left(1-x^2\right)^{1/4}$$

$$= \frac{1}{n}\left[\sqrt{\frac{\pi}{2}}p'_m\left(w^2_{n,B},x\right)w_{n,B}\left(x\right)+O\left(n^{1-\tau_1/6}\right)\right]\psi_n\left(x\right)\left(1-x^2\right)^{1/4},$$

and then the result follows.                                                    □

Next, we turn to asymptotics of orthonormal polynomials in the complex plane. In both lemmas below, we continue to assume and use (13.3) and (13.20). First we need:

**Lemma 13.8.** *For some $\tau_1 > 0, C > 0$,*
*(a)*

$$\int_{-1}^{1}|\pi_{n,m}\left(x\right)|^2e^{-2nQ_n(x)}dx \leq Cn^{-\tau_1/2}. \tag{13.36}$$

*(b) For $x = \cos\theta$, $\theta \in [0,\pi]$, let*

$$\Psi_{n,m}\left(x\right) = \sqrt{\frac{2}{\pi}}\cos\left((m-n)\theta+n\pi\int_x^1\sigma_n\left(t\right)dt+\frac{\theta}{2}-\frac{\pi}{4}\right). \tag{13.37}$$

*Then*

$$\int_{-1}^{1}\left|p_{n,m}\left(x\right)\left(1-x^2\right)^{1/4}-\Psi_{n,m}\left(x\right)e^{nQ_n(x)}\right|^2e^{-2nQ_n(x)}dx \leq Cn^{-\tau_1/2}. \tag{13.38}$$

*Proof.* (a) By (13.20), (13.21), and as $e^{-2nQ_n} \leq w_{n,B}$,

$$\int_{|x|\leq 1-n^{-\tau_1}} |\pi_{n,m}(x)|^2 e^{-2nQ_n(x)} dx \leq Cn^{-2\tau_1/3}.$$

Next, by (13.2) and our bounds (13.25) and (13.26) on the orthonormal polynomials,

$$\int_{1\geq |x|\geq 1-n^{-\tau_1}} |\pi_{n,m}(x)|^2 e^{-2nQ_n(x)} dx$$

$$\leq 2 \int_{1\geq |x|\geq 1-n^{-\tau_1}} \left( p_{n,m}^2(x) e^{-2nQ_n(x)} \right.$$

$$\left. + \left( \frac{\gamma_{n,m}}{\gamma_m(w_{n,B}^2)} \right)^2 p_m^2(w_{n,B}^2,x) w_{n,B}^2(x) \right) dx$$

$$\leq C \int_{1\geq |x|\geq 1-n^{-\tau_1}} \frac{dx}{\sqrt{1-x^2}} \leq Cn^{-\tau_1/2}.$$

(b) This is similar to (a): for $|x| \leq 1 - n^{-\tau_1/3}$, (13.4) gives

$$\left| p_{n,m}(x) e^{-nQ_n(x)} (1-x^2)^{1/4} - \Psi_{n,m}(x) \right| \leq Cn^{-\tau_1/3}.$$

For the tail integrals near $\pm 1$, we can use our bounds on the orthonormal polynomials and the bound $\sqrt{\frac{2}{\pi}}$ on $|\Psi_{n,m}|$.    $\square$

Next, we use the lemma above to estimate $\pi_{n,m}$ in the complex plane, using standard methods: recall that we defined $\phi(z) = z + \sqrt{z^2 - 1}$ in (12.22).

*Proof of Theorem 13.2(c).* Let $\pi_{n,m}$ be as in Lemma 13.7, and

$$R_{n,m}(z) = z^m \pi_{n,m}\left( \frac{1}{2}\left( z + \frac{1}{z} \right) \right).$$

Recall that $F_n$ is given by (12.14). By Cauchy's integral formula, for $|z| < 1$,

$$\left| R_{n,m}^2(z) D^2(F_n;z) \right| = \left| \frac{1}{2\pi i} \int_{|t|=1} \frac{R_{n,m}^2(t) D^2(F_n;t)}{t-z} dt \right|$$

$$\leq \frac{1}{1-|z|} \frac{1}{2\pi} \int_{-\pi}^{\pi} \left| R_{n,m}^2(e^{is}) D^2(F_n;e^{is}) \right| ds$$

$$= \frac{1}{1-|z|} \frac{1}{2\pi} \int_{-\pi}^{\pi} \left| \pi_{n,m}^2(\cos s) F_n(s) \right| ds$$

$$= \frac{1}{1-|z|} \frac{1}{\pi} \int_{-1}^{1} |\pi_{n,m}(t)|^2 e^{-2nQ_n(t)} dt$$

$$\leq \frac{Cn^{-\tau_1/2}}{1-|z|},$$

by Lemma 13.8(a). Setting $z = \phi(u)^{-1}$, where $u$ lies outside $[-1, 1]$, we have $\frac{1}{2}\left(z + \frac{1}{z}\right) = u$, so

$$\left|\phi(u)^{-m} \pi_{n,m}(u) D\left(F_n; \phi(u)^{-1}\right)\right|^2 \leq \frac{Cn^{-\tau_1/2}}{1 - \left|\phi(u)^{-1}\right|}.$$

Then for $u \in \mathbb{C} \setminus [-1, 1]$,

$$\left|\frac{\pi_{n,m}(u)}{\phi(u)^m} D\left(F_n; \phi(u)^{-1}\right)\right| \leq \frac{Cn^{-\tau_1/4}}{\left(1 - \left|\phi(u)^{-1}\right|\right)^{1/2}}. \tag{13.39}$$

Next, our definition (13.20) of $\pi_{n,m}$ gives

$$p_{n,m}(u) / \left\{ \frac{1}{\sqrt{2\pi}} \phi(u)^m D^{-1}\left(F_n; \phi(u)^{-1}\right)\right\}$$

$$= \pi_{n,m}(u) / \left\{ \frac{1}{\sqrt{2\pi}} \phi(u)^m D^{-1}\left(F_n; \phi(u)^{-1}\right)\right\}$$

$$+ \frac{\gamma_{n,m}}{\gamma_m\left(w_{n,B}^2\right)} p_m\left(w_{n,B}^2, u\right) /$$

$$\left\{ \frac{1}{\sqrt{\pi}} \phi(u)^m D^{-2}\left(\check{w}_{n,B}; \phi(u)^{-1}\right)\left(1 - \phi(u)^{-2}\right)^{-1/2}\right\} \Psi$$

where

$$\Psi = D\left(F_n; \phi(u)^{-1}\right) D^{-2}\left(\check{w}_{n,B}; \phi(u)^{-1}\right)\left(\frac{1 - \phi(u)^{-2}}{2}\right)^{-1/2}.$$

Using (13.39), (13.30), and Lemma 12.3, this gives for $u$ outside $[-1, 1]$,

$$p_{n,m}(u) / \left\{ \frac{1}{\sqrt{2\pi}} \phi(u)^m D^{-1}\left(F_n; \phi(u)^{-1}\right)\right\}$$

$$= O\left(\frac{n^{-\tau_1/4}}{\left(1 - \left|\phi(u)^{-1}\right|\right)^{1/2}}\right)$$

$$+ \left(1 + O\left(n^{-C}\right)\right)\left(1 + O\left(|\phi(u)|^{2q_n - 2m - 2}\right)\right)\Psi. \tag{13.40}$$

Here our $S_n$ in Theorem 11.1 has degree $2q_n \leq 2n - n^{1/3-2\tau_1}$, while $|m - n| \leq n^{1/3-\varepsilon}$, so $2q_n - 2m - 2 \leq -n^{1/3-2\tau_1} + 2n^{1/3-\varepsilon} \leq -\frac{1}{2}n^{1/3-2\tau_1}$, if $n$ is large enough and $\tau_1 < \varepsilon/4$, as we may assume. Also, standard estimates give

$$|\phi(u)| \geq 1 + C \, \text{dist}(u, [-1, 1])^{1/2}$$

so for $\text{dist}(u, [-1, 1]) \geq n^{-\tau_1/2}$,

$$|\phi(u)|^{2q_n - 2m - 2} \leq \exp\left(-Cn^{-\tau_1/2}n^{1/3-2\tau_1}\right) \leq n^{-\tau_1},$$

for large enough $n$. Finally, recall [25, p. 425, eqn. (26)] that $D\left(\sin^2 \cdot; z\right) = \frac{1-z^2}{2}$, so

$$\Psi = D\left(e^{-2n\breve{Q}_n}\breve{w}_{n,B}^{-2}; \phi(u)^{-1}\right) = D\left(\breve{\psi}_n^2; \phi(u)^{-1}\right)$$

$$= \exp\left(\frac{1}{4\pi}\int_{-\pi}^{\pi}\frac{e^{it} + \phi(u)^{-1}}{e^{it} - \phi(u)^{-1}}\log\psi_n^{-2}(\cos t)\,dt\right)$$

$$= \exp\left(O\left(\frac{1}{1 - |\phi(u)|^{-1}}\right)\int_{-\pi}^{\pi}|\log\psi_n(\cos t)|\,dt\right)$$

$$= \exp\left(O\left(\frac{n^{-\tau_1}}{1 - |\phi(u)|^{-1}}\right)\right),$$

by (11.7). Substituting all these estimates into (13.40) gives for some $\tau_1 > 0$ and $\text{dist}(u, [-1, 1]) \geq n^{-\tau_1}$,

$$p_{n,m}(u) \bigg/ \left\{\frac{1}{\sqrt{2\pi}}\phi(u)^m D^{-1}\left(F_n; \phi(u)^{-1}\right)\right\}$$

$$= 1 + O(n^{-\tau_1}).$$

$\square$

*Proof of Theorem 13.3.* From Lemmas 12.1(d) and 12.2(a),

$$\pi\lambda_n^{-1}\left(w_{n,B}^2, x\right)w_{n,B}^2(x)\left(1 - x^2\right)^{1/2}$$

$$= n\pi\sqrt{1 - x^2}\sigma_{Q_n}(x) + \frac{1}{\pi}PV\int_{-1}^{1}\frac{[\psi_n'(s)/\psi_n(s)]\sqrt{1 - s^2}}{s - x}ds$$

$$+ \frac{1}{2\sqrt{1 - x^2}}\sin\left((2n - 1)\theta + 2\Gamma(f_n; \theta)\right).$$

Here for some $\tau_1 > 0$ and $|x| \leq 1 - n^{-\tau_1}$, Theorem 11.1(l) gives

$$\frac{1}{\pi}PV\int_{-1}^{1}\frac{[\psi_n'(s)/\psi_n(s)]\sqrt{1 - s^2}}{s - x}ds = O\left(n^{1-\tau_1/3}\right)$$

so from Lemma 13.7(d), and (11.3) for $|x| \le 1 - n^{-\tau_1/48}$

$$\pi \lambda_n^{-1} \left( e^{-2nQ_n}, x \right) e^{-2nQ_n(x)} \left( 1 - x^2 \right)^{1/2}$$

$$= \pi \lambda_n^{-1} \left( w_{n,B}^2, x \right) w_{n,B}^2 (x) \left( 1 - x^2 \right)^{1/2} + O \left( n^{1-\tau_1/3} \right)$$

$$= n\pi \sqrt{1 - x^2} \sigma_{Q_n} (x) + O \left( n^{1-\tau_1/3} \right).$$

Then (13.7) follows, with a smaller $\tau_1$.                                                                         □

*Proof of Theorem 13.4.* The first relation in (13.8) follows directly from Theorem 13.1 and the identity $A_{n,m} = \frac{\gamma_{n,m}}{\gamma_{n,m+1}}$. For the second, namely for

$$B_{n,m} = \int_{I_n} x p_{n,m}^2 (x) \, e^{-2nQ_n(x)} dx, \tag{13.41}$$

we use an approach inspired by the Riemann-Lebesgue lemma. By the restricted range inequality Theorem 4.2(c), with $T = 1, S = n^{1/3}, R = (\log n)^{2/3}$,

$$\left| \int_{I_n \setminus [-1-\left(\frac{\log n}{n}\right)^{2/3}, 1+\left(\frac{\log n}{n}\right)^{2/3}]} x p_{n,m}^2 (x) \, e^{-2nQ_n(x)} dx \right| \le n^{-C}, \tag{13.42}$$

for some $C$, while using our bounds on $p_{n,m}$, we have for any $\eta \in \left( 0, \frac{2}{3} \right)$,

$$\left| \int_{1-n^{-\eta} \le |x| \le 1 + \left(\frac{\log n}{n}\right)^{2/3}} x p_{n,m}^2 (x) \, e^{-2nQ_n(x)} dx \right| \le C n^{-\eta/2}. \tag{13.43}$$

So we need to estimate

$$I_n = \int_{|x| \le 1 - n^{-\eta}} x p_{n,m}^2 (x) \, e^{-2nQ_n(x)} dx.$$

We may assume that $\eta$ is so small that the asymptotic (13.4) in Theorem 13.2(a) is applicable in this range. Using that asymptotic, the identity $\cos^2 t = \frac{1}{2} (1 + \cos 2t)$, and the fact that $x/\sqrt{1 - x^2}$ is an odd function, we see that for some $\tau_2 > 0$, and $\theta = \arccos x$,

$$I_n = \frac{1}{\pi} \int_{|x| \le 1 - n^{-\eta}} x \cos (2m\theta + 2\Gamma (F_n; \theta)) \frac{dx}{\sqrt{1 - x^2}} + O \left( n^{-\tau_2} \right)$$

$$= \frac{1}{2\pi} \sum_{j=-1,1} \int_{\mathcal{J}_1} \cos ((2m + j) \theta + 2\Gamma (F_n; \theta)) \, d\theta + O \left( n^{-\tau_2} \right),$$

$$= \frac{1}{2\pi} \sum_{j=-1,1} I_{n,j} + O \left( n^{-\tau_2} \right), \tag{13.44}$$

say, where $\mathcal{J}_1 = \{\theta : |\cos\theta| \leq 1 - n^{-\eta}\}$. Fix $j = \pm 1$, and let

$$h_n(\theta) = \frac{2m+j}{2n\pi}\theta + \frac{1}{n\pi}\Gamma(F_n;\theta)$$

$$= \frac{m-n+\frac{j}{2}}{n}\frac{\theta}{\pi} + \int_{\cos\theta}^{1}\sigma_{Q_n}(t)\,dt + \frac{\theta}{2n\pi} - \frac{1}{4n},$$

recall (12.15). Note that

$$h_n(0) = -\frac{1}{4n} \text{ and } h_n(\pi) = 1 + \frac{m-n}{n} + O\left(\frac{1}{n}\right) \leq 1 + O\left(n^{-2/3-\varepsilon}\right).$$

Also

$$h_n'(\theta) = \frac{m-n+\frac{j}{2}}{n\pi} + \sigma_{Q_n}(\cos\theta)\sin\theta + \frac{1}{2n\pi} > 0 \qquad (13.45)$$

in $\mathcal{J}_1$ if $\eta$ is small enough. Then, letting $h_n^{[-1]}$ denote the inverse function of $h_n$, and

$$\phi_n(t) = \frac{1}{h_n'\left(h_n^{[-1]}(t)\right)},$$

we see that

$$I_{n,j} = \int_{\mathcal{J}_1}\cos\left(2n\pi h_n(\theta)\right)d\theta$$

$$= \int_{h_n(\mathcal{J}_1)}\cos\left(2n\pi t\right)\phi_n(t)\,dt.$$

Assume now that for some $C, \beta > 0$, and $t \in h_n(\mathcal{J}_1)$,

$$\max_{|s-t|\leq\frac{1}{n}}|\phi_n(s) - \phi_n(t)| \leq Cn^{-\beta}.$$

We may assume that

$$h_n(\mathcal{J}_1) = \left[\frac{J}{n}, \frac{K}{n}\right],$$

where $-1 \leq J \leq K \leq n$. (The remaining intervals of length $O\left(\frac{1}{n}\right)$ are easily estimated.) Then we can write

$$I_{n,j} = \sum_{k=J}^{K-1} \int_{k/n}^{(k+1)/n} \cos\left(2n\pi t\right) \left[\phi_n\left(t\right) - \phi_n\left(\frac{k}{n}\right)\right] dt$$

$$= \sum_{k=J}^{K-1} O\left(n^{-1-\beta}\right) = O\left(n^{-\beta}\right).$$

Together with (13.41), (13.42), and (13.43), this gives the result. It remains to prove the hypotheses on $\phi_n$. Now from (13.45) and (3.12), in $\mathcal{J}_1$

$$h'_n\left(\theta\right) \sim \sin^2 \theta,$$

so $h_n^{[-1]}$ will satisfy a Lipschitz condition of order 1, with Lipschitz constant $O(n^{2\eta})$ in $h_n(\mathcal{J}_1)$. Moreover, since $\sigma_{Q_n}$ satisfies a Lipschitz condition of order $\alpha$, (recall (3.14–3.15)), we see that $h'_n$ also does, and hence $\phi_n$ satisfies the required smoothness condition.                                                                   □

Our final result in this chapter concerns asymptotics of zeros of orthogonal polynomials. Recall the notation (13.9), (13.11).

*Proof of Theorem 13.5.* (a) Let

$$\Delta_n\left(x\right) = p_{n,n}\left(x\right) e^{-nQ_n(x)} \left(1 - x^2\right)^{1/4}, \quad n \geq 1.$$

From (13.4) of Theorem 13.2, we know that for some $\tau_1 > 0$, and uniformly for $|x| \leq 1 - n^{-\tau_1}$, with $g_n$ defined by (13.9),

$$\Delta_n\left(x\right) = \sqrt{\frac{2}{\pi}} \cos\left(n\pi g_n\left(x\right) - \frac{\pi}{2}\right) + O\left(n^{-\tau_1}\right)$$

$$= \sqrt{\frac{2}{\pi}} \sin\left(n\pi g_n\left(x\right)\right) + O\left(n^{-\tau_1}\right). \tag{13.46}$$

Choose $j = j\left(k, n\right)$ to be the closest integer to $ng_n\left(x_{kn}\right)$. If there are two, choose the larger integer $j$. Then

$$|j - ng_n\left(x_{kn}\right)| \leq \frac{1}{2},$$

so

$$\left|n\pi\left[g_n\left(y_{jn}\right) - g_n\left(x_{kn}\right)\right]\right| \leq \frac{\pi}{2}.$$

Using the inequality $|\sin u| \geq \frac{2}{\pi}|u|$, $|u| \leq \frac{\pi}{2}$, we see that

$$\frac{2}{\pi}\left|n\pi\left[g_n\left(y_{jn}\right) - g_n\left(x_{kn}\right)\right]\right| \leq \left|\sin n\pi\left[g_n\left(y_{jn}\right) - g_n\left(x_{kn}\right)\right]\right|$$

$$= \sqrt{\frac{\pi}{2}}\left|\Delta_n\left(x_{kn}\right)\right| + O\left(n^{-\tau_1}\right) = O\left(n^{-\tau_1}\right),$$

by (13.46). Thus

$$\left| g_n\left(y_{jn}\right) - g_n\left(x_{kn}\right) \right| \leq Cn^{-1-\tau_1}.$$

Since (13.10) shows that

$$g_n'\left(t\right) \sim -\sqrt{1-t^2}, \quad |t| \leq 1 - n^{-\tau_1},$$

we have

$$\left| y_{jn} - x_{kn} \right| \leq Cn^{-1-\tau_1} / \sqrt{1 - x_{kn}^2},$$

which yields (13.13) if $\tau_2$ is small enough.

Next, suppose that $x_{kn}, x_{k+1,n}$ satisfy (13.13) with the same $j$, so that

$$x_{kn} - x_{k+1,n} \leq Cn^{-1-\tau_2}.$$

We shall derive a contradiction. By Rolle's theorem, there exists $y \in (x_{k+1,n}, x_{kn})$ with $p_{n,n}'(y) = 0$. We show that this contradicts our asymptotic for $p_{n,n}'$ in (13.5). First, that asymptotic gives

$$\frac{1}{n} \sqrt{\frac{\pi}{2}} \left| p_{n,n}'\left(x_{kn}\right) \right| e^{-nQ_n(x_{kn})} \left(1 - x_{kn}^2\right)^{1/4}$$

$$= \left| -\pi \sigma_{Q_n}\left(x_{kn}\right) \cos\left(n\pi g_n\left(x_{kn}\right)\right) + Q_n'\left(x_{kn}\right) \sin\left(n\pi g_n\left(x_{kn}\right)\right) + O\left(n^{-\tau_1}\right) \right|$$

$$= \pi \sigma_{Q_n}\left(x_{kn}\right)\left(1 + O\left(n^{-\tau_1}\right)\right) + O\left(n^{-\tau_1}\right).$$

Thus using (3.12),

$$\left| p_{n,n}'\left(x_{kn}\right) \right| e^{-nQ_n(x_{kn})} \sim n\left(1 - x_{kn}^2\right)^{1/4}.$$

Then for some $z$ between $y$ and $x_{kn}$, we have

$$\left| \left(p_{n,n}'e^{-nQ_n}\right)'\left(z\right) \right| \geq Cn\left(1 - x_{kn}^2\right)^{1/4} / |y - x_{kn}| \geq Cn^{2+\tau_2}\left(1 - x_{kn}^2\right)^{1/4}. \tag{13.47}$$

Now our bounds (7.3) and the restricted range inequality Theorem 4.2 show that

$$\sup_{x \in I_n} \left| \left(1 - x^2\right) p_{n,n}\left(x\right) e^{-nQ_n(x)} \right| \leq C,$$

so by Theorem 8.1(b),

$$\sup_{x \in I_n} \left| \left[ \left(1 - x^2\right) p_{n,n}'\left(x\right) - 2x p_{n,n}\left(x\right) \right] e^{-nQ_n(x)} \right| \leq Cn.$$

Then

$$\sup_{x \in I_n} \left| \left(1 - x^2\right) p'_{n,n}(x) e^{-n Q_n(x)} \right| \leq Cn.$$

Applying Theorem 8.1(b) again in a similar fashion yields

$$\sup_{x \in I_n} \left| \left(1 - x^2\right) p''_{n,n}(x) e^{-n Q_n(x)} \right| \leq Cn^2.$$

This clearly contradicts (13.47) for $|x_{kn}| \leq 1 - n^{-\tau_2}$ and some possibly smaller $\tau_2$. Thus there is at most one $x_{kn}$ within $O\left(n^{-1-\tau_2}\right)$ of $y_{jn}$. The fact that there is actually one follows from the asymptotic (13.4) and the sign change of $\sin(n\pi g_n(t))$ at $t = y_{jn}$.

   (b) In view of the spacing (13.13) and that each $y_{jn}$ is "close" to exactly one $x_{kn}$, (a) gives

$$\sigma_n(x_{kn})(x_{kn} - x_{k+1,n}) = \sigma_n(x_{kn})\left(y_{jn} - y_{j+1,n}\right) + O\left(n^{-1-\tau_1}\right), \tag{13.48}$$

uniformly for $|x_{kn}| \leq 1 - n^{-\tau_1}$. Finally, for some $t \in \left(y_{j+1,n}, y_{jn}\right)$, our definition (13.11) of $y_{jn}$ gives

$$
\begin{aligned}
\frac{1}{n} &= g'_n(t)\left(y_{j+1,n} - y_{jn}\right) \\
&= \left(-\sigma_{Q_n}(t) - \frac{1}{2\pi n\sqrt{1 - t^2}}\right)\left(y_{j+1,n} - y_{jn}\right) \\
&= \left(\sigma_{Q_n}(x_{kn}) + O\left(|t - x_{kn}|^\alpha\right) + O\left(\frac{1}{n\sqrt{1 - y_{jn}^2}}\right)\right)\left(y_{jn} - y_{j+1,n}\right) \\
&= \sigma_{Q_n}(x_{kn})\left(y_{jn} - y_{j+1,n}\right) + O\left(\frac{1}{n\sqrt{1 - y_{jn}^2}}\right)^{1+\alpha},
\end{aligned}
$$

by (13.12). Substituting into (13.48) gives the result.                                              □

# Chapter 14
# Further Bounds

Recall that the fundamental polynomials $\{\ell_{jn}\}$ of Lagrange interpolation at the zeros of $p_{n,n}$ are given by

$$\ell_{jn}(x) = \frac{p_{n,n}(x)}{p'_{n,n}(x_{jn})(x - x_{jn})} = \frac{K_n(x, x_{jn})}{K_n(x_{jn}, x_{jn})}. \tag{14.1}$$

Throughout, we assume that $\{Q_n\} \in \mathcal{Q}$ and that $I_0$ and $r^* > 1$ are as in Definition 1.1. We prove:

**Theorem 14.1.** *Uniformly for* $1 \le j \le n$,

$$\sup_{x \in I_n} \left| \ell_{jn}(x) \right| e^{-nQ_n(x)} / e^{-nQ_n(x_{jn})} \le C. \tag{14.2}$$

We shall deduce:

**Theorem 14.2.** *Let* $A > 0$. *For* $n \ge 1$, *choose* $m = m(n) \ge 1$ *with*

$$|m(n) - n| \le An^{1/3}. \tag{14.3}$$

*Let* $\{\hat{x}_{jm}\}$ *denote the zeros of* $p_{n,m}$ *in decreasing order. Uniformly for* $n \ge 1$, $1 \le j \le m - 1$,
*(a) for* $x \in \left[\hat{x}_{j+1,m}, \hat{x}_{jm}\right]$,

$$\left| p_{n,m}(x) e^{-nQ_n(x)} \right|$$

$$\sim n \max \left\{1 - \left|\hat{x}_{jm}\right|, n^{-2/3}\right\}^{1/4} \min \left\{\left|x - \hat{x}_{jm}\right|, \left|x - \hat{x}_{j+1,m}\right|\right\}. \tag{14.4}$$

E. Levin, D.S. Lubinsky, *Bounds and Asymptotics for Orthogonal Polynomials
for Varying Weights*, SpringerBriefs in Mathematics,
https://doi.org/10.1007/978-3-319-72947-3_14

*(b)*

$$\left| p'_{n,m} \left( \hat{x}_{jm} \right) \right| e^{-nQ_n(\hat{x}_{jm})} \sim n \max \left\{ 1 - \left| \hat{x}_{jm} \right|, n^{-2/3} \right\}^{1/4} \qquad (14.5)$$

*and*

$$\left| p_{n,m-1} \left( \hat{x}_{jm} \right) \right| e^{-nQ_n(\hat{x}_{jm})} \sim \max \left\{ 1 - \left| \hat{x}_{jm} \right|, n^{-2/3} \right\}^{1/4}. \qquad (14.6)$$

*(c)*

$$\hat{x}_{jm} - \hat{x}_{j+1,m} \sim \frac{1}{n} \max \left\{ 1 - \left| \hat{x}_{jm} \right|, n^{-2/3} \right\}^{-1/2}. \qquad (14.7)$$

*(d)*

$$\left\| p_{n,m} e^{-nQ_n} \right\|_{L_\infty(I_n)} \sim n^{1/6}. \qquad (14.8)$$

This chapter is structured as follows: we first estimate the fundamental polynomials. Then we prove Theorem 14.1, followed by Theorem 14.2 for the special case $m(n) = n$, $n \geq 1$. Then we deduce the general case. Recall from (7.4) that

$$A_n(x) = 2n \int_{I_n} p_{n,n}^2(t) \, \bar{Q}_n(x,t) \, e^{-2nQ_n(t)} dt.$$

**Lemma 14.3.** *Let* $1 < r^\# < r^*$.
*(a) Uniformly for* $n \geq 1$ *and* $x \in \left[ -r^\#, r^\# \right]$,

$$A_n(x) \sim n. \qquad (14.9)$$

*(b) For* $|x| \leq 1$, *and* $1 \leq j \leq n$,

$$\left| \ell_{jn}(x) \right| e^{-nQ_n(x)} / e^{-nQ_n(x_{jn})} \leq C \left\{ \frac{\max \left\{ 1 - |x|, n^{-2/3} \right\}}{\max \left\{ 1 - \left| x_{jn} \right|, n^{-2/3} \right\}} \right\}^{1/4}. \qquad (14.10)$$

*(c) For* $|x| \leq 1$, *and* $1 \leq j \leq n$,

$$\left| \ell_{jn}(x) \right| e^{-nQ_n(x)} / e^{-nQ_n(x_{jn})} \leq C \frac{\left( 1 - x^2 \right)^{-1/4} \left( \max \left\{ 1 - \left| x_{jn} \right|, n^{-2/3} \right\} \right)^{-1/4}}{n \left| x - x_{jn} \right|}. \qquad (14.11)$$

*Proof.* (a) Let $0 \le x \le r^\#$. Assume that $\eta$ is so small that $[1 - 2\eta, 1 + 2\eta] \subset I_0$. By Theorem 7.1, and our assumptions in Definition 1.1,

$$\int_{-r^*}^{r^*} p_{n,n}^2(t) \, \bar{Q}_n(x, t) \, e^{-2nQ_n(t)} \, dt$$

$$\le C \int_{-r^*}^{r^*} \bar{Q}_n(x, t) \, \frac{dt}{\sqrt{|1 - t^2|}}$$

$$\le C \left[ \int_{[-r^*, r^*] \setminus [1-\eta, 1+\eta]} |x - t|^{\alpha - 1} \, dt + \int_{[1-\eta, 1+\eta]} \frac{\bar{Q}_n(x, t)}{\sqrt{|1 - t|}} \, dt \right]$$

$$\le C \left[ \int_{-4}^{4} |s|^{\alpha - 1} \, ds + \int_{[1-\eta, 1+\eta]} \frac{\bar{Q}_n(x, t)}{\sqrt{|1 - t|}} \, dt \right]. \tag{14.12}$$

Here if $x \notin [1 - 2\eta, 1 + 2\eta]$, we can estimate the second integral by

$$\frac{1}{\eta} \int_{1-\eta}^{1+\eta} \left[ |Q_n'(t)| + |Q_n'(x)| \right] \frac{dt}{\sqrt{|1 - t^2|}} \le C,$$

using uniform boundedness of $\{Q_n'\}$ in $[-r^*, r^*]$. If $x \in [1 - 2\eta, 1 + 2\eta]$, we instead estimate the second integral by using the substitution $(1 - t) = s(1 - x)$:

$$C \int_{1-\eta}^{1+\eta} |x - t|^{\alpha_1 - 1} \frac{dt}{\sqrt{|1 - t|}}$$

$$= C |1 - x|^{\alpha_1 - \frac{1}{2}} \int_{-\frac{\eta}{|1-x|}}^{\frac{\eta}{|1-x|}} |1 - s|^{\alpha_1 - 1} \frac{ds}{\sqrt{|s|}}$$

$$\le C |1 - x|^{\alpha_1 - \frac{1}{2}} \left\{ 1 + \left( \frac{\eta}{|1 - x|} \right)^{\alpha_1 - \frac{1}{2}} \right\} \le C,$$

as $\alpha_1 \ge \frac{1}{2}$. Thus in summary, for $0 \le x \le r^\#$, together with (14.12), this gives

$$\int_{-r^*}^{r^*} p_{n,n}^2(t) \, \bar{Q}_n(x, t) \, e^{-2nQ_n(t)} \, dt \le C.$$

Next, for such $x$,

$$\int_{I_n \setminus [-r^*, r^*]} p_{n,n}^2(t) \, \bar{Q}_n(x, t) \, e^{-2nQ_n(t)} \, dt$$

$$\le C \left[ \int_{I_n \setminus [-r^*, r^*]} p_{n,n}^2(t) \, |Q_n'(t)| \, e^{-2nQ_n(t)} \, dt + |Q_n'(x)| \right] \le C,$$

by (7.9), and using uniform boundedness of $\{Q'_n\}$ in $[-r^*, r^*]$. Thus

$$\sup_{x \in [-r^\#, r^\#]} A_n(x) \leq Cn.$$

Now we establish a matching lower bound. Let us suppose that $x \geq t_n$, where $Q'_n(t_n) = 0$, so that $Q'_n(x) \geq 0 \geq Q'_n(t)$ for $t \leq t_n$. Then as $\bar{Q}_n(x, t) \geq 0$, we have

$$A_n(x) \geq Cn \int_{c_n}^{t_n} p_{n,n}^2(t) \frac{|Q'_n(x)| + |Q'_n(t)|}{|x - t|} e^{-2nQ_n(t)} dt$$

$$\geq \frac{Cn}{2r^\#} \int_{-r^\#}^{t_n} p_{n,n}^2(t) |Q'_n(t)| e^{-2nQ_n(t)} dt. \qquad (14.13)$$

Next,

$$2n \int_{c_n}^{-r^\#} p_{n,n}^2(t) |Q'_n(t)| e^{-2nQ_n(t)} dt$$

$$= \int_{c_n}^{-r^\#} p_{n,n}^2(t) \frac{d}{dt} \left\{ e^{-2nQ_n(t)} \right\} dt$$

$$= \left[ p_{n,n}^2(t) e^{-2nQ_n(t)} \right]_{t=c_n}^{t=-r^\#} - 2 \int_{c_n}^{-r^\#} p'_{n,n}(t) p_{n,n}(t) e^{-2nQ_n(t)} dt$$

$$\leq p_{n,n}^2(-r^\#) e^{-2nQ_n(-r^\#)} + C_1 e^{-C_2 n} \int_{-1}^{1} |p'_{n,n}(t) p_{n,n}(t)| e^{-2nQ_n(t)} dt$$

$$\leq C_1 e^{-C_2 n} + C_1 e^{-C_2 n} n^{C_3},$$

by a straightforward application of our restricted range inequality Theorem 4.2(c), our Markov-Bernstein inequality Theorem 8.1, and our bounds on $\{p_{n,n}\}$. Then, using Lemma 7.3(b),

$$\int_{-r^\#}^{t_n} p_{n,n}^2(t) |Q'_n(t)| e^{-2nQ_n(t)} dt$$

$$= \int_{c_n}^{t_n} p_{n,n}^2(t) |Q'_n(t)| e^{-2nQ_n(t)} dt + O\left(e^{-C_3 n}\right)$$

$$= \frac{1}{2} \int_{I_n} p_{n,n}^2(t) |Q'_n(t)| e^{-2nQ_n(t)} dt + O\left(e^{-C_3 n}\right). \qquad (14.14)$$

Next, by Lemma 7.3(a),

$$1 + \frac{1}{2n} \leq \int_{I_n} p_{n,n}^2(t) \left| t Q_n'(t) \right| e^{-2nQ_n(t)} dt$$

$$\leq r^{\#} \int_{-r^{\#}}^{r^{\#}} p_{n,n}^2(t) \left| Q_n'(t) \right| e^{-2nQ_n(t)} dt$$

$$+ \frac{1}{2n} \int_{I_n \setminus [-r^{\#}, r^{\#}]} \left| t p_{n,n}^2(t) \right| \left| \frac{d}{dt} \left( e^{-2nQ_n(t)} \right) \right| dt$$

$$\leq r^{\#} \int_{I_n} p_{n,n}^2(t) \left| Q_n'(t) \right| e^{-2nQ_n(t)} dt + O\left( e^{-C_2 n} \right),$$

much as above. So together with (14.14), this gives

$$\int_{-r^{\#}}^{t_n} p_{n,n}^2(t) \left| Q_n'(t) \right| e^{-2nQ_n(t)} dt \geq C_4,$$

so from (14.13),

$$A_n(x) \geq Cn.$$

(b) Using Cauchy-Schwarz and our bounds on $K_n = \lambda_{n,2}^{-1}$ from Theorem 5.1, as well as the fact that all $\left| x_{jn} \right| \leq 1 + \frac{C_2}{n}$,

$$\left| \ell_{jn}(x) \right| e^{-nQ_n(x)} / e^{-nQ_n(x_{jn})} = \left| \frac{K_n(x, x_{jn}) e^{-nQ_n(x)}}{K_n(x_{jn}, x_{jn}) e^{-nQ_n(x_{jn})}} \right|$$

$$\leq \left\{ \frac{K_n(x, x) e^{-2nQ_n(x)}}{K_n(x_{jn}, x_{jn}) e^{-2nQ_n(x_{jn})}} \right\}^{1/2}$$

$$\leq C \left\{ \frac{\max\{1 - |x|, n^{-2/3}\}}{\max\{1 - |x_{jn}|, n^{-2/3}\}} \right\}^{1/4}.$$

(c) Recall first that

$$K_n(x_{jn}, x_{jn}) = \frac{\gamma_{n,n-1}}{\gamma_{n,n}} p_{n,n}'(x_{jn}) p_{n,n-1}(x_{jn})$$

$$= p_{n,n}'(x_{jn})^2 / A_n(x_{jn}), \qquad (14.15)$$

by (7.5). Next, using the Christoffel-Darboux formula and then (7.5) again,

$$\left| \ell_{jn} (x) \right| e^{-nQ_n(x)} / e^{-nQ_n(x_{jn})}$$

$$= \frac{\gamma_{n,n-1}}{\gamma_{n,n}} \frac{\left| p_{n,n-1} (x_{jn}) \right| \left| p_{n,n} (x) \right| e^{-nQ_n(x)}}{\left| x - x_{jn} \right| K_n (x_{jn}, x_{jn}) e^{-nQ_n(x_{jn})}}$$

$$= \frac{\left| p'_{n,n} (x_{jn}) \right| \left| p_{n,n} (x) \right| e^{-nQ_n(x)}}{A_n (x_{jn}) \left| x - x_{jn} \right| K_n (x_{jn}, x_{jn}) e^{-nQ_n(x_{jn})}}$$

$$= \frac{\left| p_{n,n} (x) \right| e^{-nQ_n(x)}}{\left| x - x_{jn} \right| \left( A_n (x_{jn}) K_n (x_{jn}, x_{jn}) \right)^{1/2} e^{-nQ_n(x_{jn})}}$$

$$\leq C \frac{\left( 1 - x^2 \right)^{-1/4} \left( \max \left\{ 1 - \left| x_{jn} \right|, n^{-2/3} \right\} \right)^{-1/4}}{n \left| x - x_{jn} \right|},$$

by (14.15), (a) of this lemma, (5.3), and our bounds on $p_{n,n}$ in Theorem 7.1.   $\square$

*Proof of Theorem* 14.1. If first, $\max \left\{ 1 - \left| x \right|, n^{-2/3} \right\} \leq 2 \max \left\{ 1 - \left| x_{jn} \right|, n^{-2/3} \right\}$, then our first bound (14.10) gives

$$\left| \ell_{jn} (x) \right| e^{-nQ_n(x)} / e^{-nQ_n(x_{jn})} < C.$$

Otherwise, $1 - \left| x \right| \geq 2 \left( 1 - \left| x_{jn} \right| \right)$, and $1 - \left| x \right| \geq 2n^{-2/3}$, so

$$\left| x - x_{jn} \right| \geq \left| \left| x \right| - \left| x_{jn} \right| \right|$$

$$= \left| \left( 1 - \left| x \right| \right) - \left( 1 - \left| x_{jn} \right| \right) \right| \geq \frac{1}{2} \left( 1 - \left| x \right| \right),$$

so from the second bound (14.11),

$$\left| \ell_{jn} (x) \right| e^{-nQ_n(x)} / e^{-nQ_n(x_{jn})}$$

$$\leq \frac{1}{n} \left( 1 - \left| x \right| \right)^{-5/4} \left( \max \left\{ 1 - \left| x_{jn} \right|, n^{-2/3} \right\} \right)^{-1/4}$$

$$\leq \frac{C}{n} \left( n^{\frac{2}{3}} \right)^{\frac{5}{4} + \frac{1}{4}} = C.$$

Thus

$$\sup_{|x| \leq 1} \left| \ell_{jn} (x) \right| e^{-nQ_n(x)} / e^{-nQ_n(x_{jn})} \leq C.$$

Then our restricted range inequality Theorem 4.2 gives the rest.                □

*Proof of Theorem 14.2(a), (b) for* $m = m(n) = n$. Let $1 < r^\# < r^*$. From Theorem 14.1 and (14.15), for all $|x| \le r^\#$, $1 \le j \le n$,

$$|p_{n,n}(x)| e^{-nQ_n(x)} \le C |x - x_{jn}| \left| p'_{n,n}(x_{jn}) e^{-nQ_n(x_{jn})} \right|$$

$$= C |x - x_{jn}| \left[ K_n(x_{jn}, x_{jn}) A_n(x_{jn}) \right]^{1/2} e^{-nQ_n(x_{jn})}$$

$$\le Cn |x - x_{jn}| \max \left\{ 1 - |x_{jn}|, n^{-2/3} \right\}^{1/4},$$

recall (5.2). Then the upper bound implicit in (14.4) follows, provided we also use

$$\max \left\{ 1 - |x_{jn}|, n^{-2/3} \right\} \sim \max \left\{ 1 - |x_{j+1,n}|, n^{-2/3} \right\},$$

which was proved at (6.5). We turn to the matching lower bound. From Lemma 6.3, for $x \in [x_{j+1,n}, x_{jn}]$,

$$\ell_{jn}(x) e^{-nQ_n(x)} e^{nQ_n(x_{jn})} + \ell_{j+1,n}(x) e^{-nQ_n(x)} e^{nQ_n(x_{j+1,n})} \ge 1,$$

so

$$p_{n,n}(x) e^{-nQ_n(x)} \left\{ \frac{1}{(x - x_{jn}) p'_{n,n}(x_{jn}) e^{-nQ_n(x_{jn})}} \right.$$

$$\left. + \frac{1}{(x - x_{j+1,n}) p'_{n,n}(x_{j+1,n}) e^{-nQ_n(x_{j+1,n})}} \right\} \ge 1. \tag{14.16}$$

Now from (14.15), followed by (14.9) and Theorem 5.1, and for $k = j, j+1$,

$$\left| p'_{n,n}(x_{kn}) e^{-nQ_n(x_{kn})} \right| = \left( A_n(x_{kn}) K_n(x_{kn}, x_{kn}) e^{-2nQ_n(x_{kn})} \right)^{1/2}$$

$$\sim n \max \left\{ 1 - |x_{kn}|, n^{-2/3} \right\}^{1/4}$$

$$\sim n \max \left\{ 1 - |x_{jn}|, n^{-2/3} \right\}^{1/4}$$

by (6.5). Then (14.16) gives the matching lower bound

$$\left| p_{n,n}(x) e^{-nQ_n(x)} \right|$$

$$\ge Cn \max \left\{ 1 - |x_{in}|, n^{-2/3} \right\}^{1/4} \min \left\{ |x - x_{jn}|, |x - x_{j+1,n}| \right\}.$$

So we have (14.4) for $m = n$. This also gives (14.5) (let $x \to x_{jn}$) and then (14.6) follows from (7.5).                                                              $\square$

*Proof of Theorem 14.2(c), (d) for $m = m(n) = n$.* (c) The upper bound was proved in Theorem 6.1(b), so we need to only prove the lower bound. Now for some $\xi$ between $x_{jn}$ and $x_{j+1,n}$,

$$1 = \ell_{jn}\left(x_{jn}\right) e^{-nQ_n(x_{jn})} e^{nQ_n(x_{jn})} - \ell_{jn}\left(x_{j+1,n}\right) e^{-nQ_n(x_{j+1,n})} e^{nQ_n(x_{jn})}$$

$$= \left(x_{jn} - x_{j+1,n}\right) \frac{d}{dx}\left\{\ell_{jn}\left(x\right) e^{-nQ_n(x)} e^{nQ_n(x_{jn})}\right\}_{|x=\xi}$$

$$\leq C\left(x_{jn} - x_{j+1,n}\right) n \max\left\{1 - |\xi|, n^{-2/3}\right\}^{1/2},$$

by Theorem 8.1(a) and our bound above on $\ell_{jn}$. Using (6.5), we continue this as

$$1 \leq C\left(x_{jn} - x_{j+1,n}\right) n \max\left\{1 - |x_{jn}|, n^{-2/3}\right\}^{1/2}.$$

(d) The upper bound implicit in (14.8) follows from Theorem 7.1(b). For the lower bound, we use (a) of this theorem with $x = \frac{1}{2}\left(x_{1n} + x_{2n}\right), j = 1$, to obtain

$$\left\|P_{n,n}e^{-nQ_n}\right\|_{L_\infty(I_n)} \geq Cn \max\left\{|1 - |x_{1n}||, n^{-2/3}\right\}^{1/4} |x_{1n} - x_{2n}|$$

$$\geq C \max\left\{|1 - |x_{1n}||, n^{-2/3}\right\}^{-1/4} \geq Cn^{1/6},$$

by (c) and (6.1).                                                              $\square$

*Proof of Theorem 14.2 for the General Case $m = m(n)$ Satisfying (14.3).* As in the general case of Theorem 7.1, let

$$r_n = m/n$$

and for $x \in L_{n,r_n}\left(I_n\right) =: I_m^\#$,

$$Q_m^\#\left(x\right) = \frac{n}{m} Q_n\left(L_{n,r_n}^{[-1]}\left(x\right)\right) = r_n^{-1} Q_n\left(L_{n,r_n}^{[-1]}\left(x\right)\right),$$

so that

$$m Q_m^\#\left(x\right) = n Q_n\left(t\right), \quad t = L_{n,r_n}^{[-1]}\left(x\right).$$

If $p_m^\#$ is the orthonormal polynomial of degree $m$ for $e^{-2mQ_m^\#}$, then we showed in the proof of the general case of Theorem 7.1 that

$$p_m^\#(x) = \sqrt{\delta_{n,r_n}} p_{n,m}\left(L_{n,r_n}^{[-1]}(x)\right).$$

Moreover, we have (7.21). It is then straightforward to transfer the estimates from the special case $m(n) = n$ to the general case above.                               $\square$

# Chapter 15
# Universality Limits and Entropy Integrals

The theory of random matrices goes back at least to John Wishart, but it was the pathbreaking work of Eugene Wigner in modeling nuclei of heavy atoms that brought them to real prominence. He speculated that the spacings between the lines in the spectrum of such a nucleus resemble spacings between the eigenvalues of a random Hermitian matrix. Wigner and Freeman Dyson, amongst others, made major initial contributions to a topic that remains a very active area of research [13].

Wigner considered a probability distribution that after a transformation, becomes a probability distribution $\mathbb{P}^{(n)}$ on the eigenvalues $\lambda_1 \leq \lambda_2 \leq \cdots \leq \lambda_n$ of an $n \times n$ Hermitian matrix $M$. The probability density function for $\mathbb{P}^{(n)}$, which we denote by $\mathcal{P}^{(n)}$, takes the form

$$\mathcal{P}^{(n)}(\lambda_1, \lambda_2, \ldots, \lambda_n) = \frac{1}{Z_n}\left(\prod_{1 \leq i < j \leq n}(\lambda_i - \lambda_j)^2\right)\left(\prod_{j=1}^{n} e^{-2nQ_n(\lambda_j)}\right). \tag{15.1}$$

Here $Z_n$ is a normalizing constant. In the classical Gaussian Unitary Ensemble of Wigner, $Q_n(x) = \frac{1}{2}x^2$ and one wants to consider various phenomena as $n \to \infty$.

One important quantity that permits asymptotics as $n \to \infty$ is the $m$-point correlation function [9, p. 112]:

$$\mathcal{R}_m^{(n)}(\lambda_1, \lambda_2, \ldots, \lambda_m)$$
$$= \frac{n!}{(n-m)!} \int \cdots \int \mathcal{P}^{(n)}(\lambda_1, \lambda_2, \ldots, \lambda_n)\, d\lambda_{m+1}\, d\lambda_{m+2} \cdots d\lambda_n.$$

© The Author(s) 2018
E. Levin, D.S. Lubinsky, *Bounds and Asymptotics for Orthogonal Polynomials for Varying Weights*, SpringerBriefs in Mathematics,
https://doi.org/10.1007/978-3-319-72947-3_15

It allows one to find the expected number of $m$-tuples of eigenvalues in a given compact set. Typically one analyzes this with $m$ fixed and $n \to \infty$, using a connection to orthogonal polynomials, first discovered by Freeman Dyson [9, p. 112]:

$$\mathcal{R}_m^{(n)}\left(\lambda_1, \lambda_2, \ldots, \lambda_m\right) = \det\left(K_n\left(\lambda_i, \lambda_j\right) e^{-nQ_n(\lambda_i)} e^{-nQ_n(\lambda_j)}\right)_{1 \le i,j \le m}$$

$$= \det\left(\tilde{K}_n\left(\lambda_i, \lambda_j\right)\right)_{1 \le i,j \le m}, \tag{15.2}$$

where we set

$$\tilde{K}_n\left(x, y\right) = K_n\left(x, y\right) e^{-nQ_n(x)} e^{-nQ_n(y)}.$$

The *universality limit in the bulk*, which describes the "microlocal" spacing of the eigenvalues, asserts that for fixed $m \ge 2$, $x$ in a suitable subset of the (common) supports of $\{e^{-2nQ_n}\}$, and real $a_1, a_2, \ldots, a_m$, we have

$$\lim_{n \to \infty} \frac{1}{\tilde{K}_n\left(x, x\right)^m} \mathcal{R}_m^{(n)}\left(x + \frac{a_1}{\tilde{K}_n\left(x, x\right)}, x + \frac{a_2}{\tilde{K}_n\left(x, x\right)}, \ldots, x + \frac{a_m}{\tilde{K}_n\left(x, x\right)}\right)$$

$$= \det\left(\frac{\sin \pi\left(a_i - a_j\right)}{\pi\left(a_i - a_j\right)}\right)_{1 \le i,j \le m}. \tag{15.3}$$

Because we are dealing with fixed size matrices, Dyson's identity (15.2) permits reduction of (15.3) to just the technical limit

$$\lim_{n \to \infty} \frac{\tilde{K}_n\left(x + \frac{a_1}{\tilde{K}_n(x,x)}, x + \frac{a_2}{\tilde{K}_n(x,x)}\right)}{\tilde{K}_n\left(x, x\right)} = \frac{\sin \pi\left(a_1 - a_2\right)}{\pi\left(a_1 - a_2\right)}.$$

One natural approach to this limit is to substitute sufficiently precise asymptotics for orthogonal polynomials into the Christoffel-Darboux formula. Indeed, this was one early approach. Undoubtedly the most spectacular advances came from the two-dimensional steepest descent method developed by Deift and Zhou [9, 12]. They made heavy use of Riemann-Hilbert problems, as well as earlier ideas of Fokas, Its, and Kitaev, and established far more than universality limits, primarily for analytic or piecewise analytic $Q_n$. Subsequently the second author developed comparison [28] and normality methods [29] that can handle fixed or varying $Q_n$ with less smoothness. Barry Simon [45, 46] and Vili Totik [53, 54] developed these ideas for fixed measures, while the authors applied these to varying weights $\{e^{-nQ}\}$ in [27].

See [1, 2, 5, 9, 10, 13, 31] for some of the surveys of this very active topic.

Here we use the elementary method of substituting asymptotics for orthonormal polynomials into the Christoffel-Darboux formula to prove:

**Theorem 15.1.** *Let* $\{Q_n\} \in \mathcal{Q}$. *There exists* $\tau > 0$ *such that as* $n \to \infty$, *uniformly for* $|x| \leq 1 - n^{-\tau}$ *and* $a, b$ *in compact subsets of the real line,*

$$\frac{\tilde{K}_n\left(x + \frac{a}{\tilde{K}_n(x,x)}, x + \frac{b}{\tilde{K}_n(x,x)}\right)}{\tilde{K}_n(x,x)} = \frac{\sin \pi (a - b)}{\pi (a - b)} + O\left(n^{-\tau}\right). \tag{15.4}$$

**Remarks.** (a) This immediately implies (15.3).

(b) Note that for weights of the form $e^{-nQ}$, where the equilibrium measure for $Q$ is supported on finitely many intervals, we proved such limits in [27] under weaker conditions on $Q'$ (namely continuity of $Q'$ and its equilibrium density) than we impose on $\{Q_n\}$. However this was without a rate, and Theorem 15.1 seems to be the first result of its type for varying $\{Q_n\}$.

We also consider fluctuations of the eigenvalues, along the lines of earlier work of Jonas Gustavsson [17] and Deng Zhang [58, 59], and the second author [30]. They showed that for $j/n$ bounded away from 0 and 1, the scaled difference between the $j$th random eigenvalue $\lambda_j$ and $j$th zero of $p_{n,n}(\cdot)$ satisfies [58, p. 1490, Thm. 1.4]

$$\frac{\lambda_j - x_{jn}}{\sqrt{\frac{\log n}{2\pi^2} \frac{1}{n\sigma_n^*(x_{jn})}}} \to N(0, 1) \tag{15.5}$$

in distribution as $n \to \infty$. Here $N(0, 1)$ is the normal distribution, that is, has probability density $\frac{1}{\sqrt{2\pi}} e^{-\frac{1}{2}t^2}, t \in (-\infty, \infty)$. (Their ordering of zeros is different.) We prove:

**Theorem 15.2.** *Let* $\{Q_n\} \in \mathcal{Q}$. *Let* $\varepsilon \in \left(0, \frac{1}{2}\right)$. *For* $n \geq 1$, *let* $\mathbb{P}^{(n)}$ *denote the probability distribution with density defined by* (15.1) *for all* $n \geq 1$. *Let* $\lambda_1, \lambda_2, \ldots, \lambda_n$ *denote the eigenvalues in increasing order. Also, let* $\{x_{jn}\}$ *denote the zeros of* $p_{n,n}$. *Then for* $j, n$ *with* $x_{jn} \in [-1 + \varepsilon, 1 - \varepsilon]$, *we have in distribution as* $n \to \infty$,

$$\frac{\lambda_j - x_{n+1-j,n}}{\sqrt{\frac{\log n}{2\pi^2} \frac{1}{n\sigma_{Q_n}(x_{n+1-j,n})}}} \to N(0, 1). \tag{15.6}$$

We note that the result holds for zeros in $[-1 + n^{-\tau_1}, 1 - n^{-\tau_1}]$ and not just $[-1 + \varepsilon, 1 - \varepsilon]$, but we are restricted by the formulation in [30], which applies only to compact subsets of $(-1, 1)$. Finally, we turn to entropy type integrals. The main example is [4, 6, 8, 26]

$$-\int \left| p_{n,n} e^{-nQ_n} \right|^2 \log p_{n,n}^2.$$

Since in our case $\log|p_{n,n}| = nQ_n+$ a smaller locally integrable term, this last integral has the same asymptotic as

$$\int \left| p_{n,n} e^{-nQ_n} \right|^2 \left| \log p_{n,n}^2 \right|.$$

We consider slightly more general integrals:

**Theorem 15.3.** *Let* $\{Q_n\} \in \mathcal{Q}$. *Let*

$$0 < \rho < 4 \text{ and } \tau \geq 0. \qquad (15.7)$$

*Define*

$$\mathcal{I}(n) = \int_{I_n} \left| p_{n,n}(x) e^{-nQ_n(x)} \right|^\rho \left| \log|p_{n,n}(x)| \right|^\tau dx. \qquad (15.8)$$

*Then as* $n \to \infty$,

$$\mathcal{I}(n)/n^\tau = (1 + o(1)) \frac{2^{\frac{3}{2}\rho} \Gamma\left(\frac{\rho+1}{2}\right)^2}{\pi^{\rho/2+1} \Gamma(\rho+1)} \int_{-1}^{1} |Q_n(t)|^\tau \left(1 - t^2\right)^{-\rho/4} dt. \qquad (15.9)$$

We note that careful tracking of estimates allows one to replace $1 + o(1)$ by $1 + o(n^{-\tau})$ for some $\tau > 0$.

*Proof of Theorem 15.1.* Rather than applying the more general results from [27], we substitute our asymptotics for $\{p_{n,n}\}$ into the Christoffel-Darboux formula: for $x \in (-1, 1)$, let

$$x_{n,a} = x + \frac{a}{\widetilde{K}_n(x,x)}, \qquad n \geq 1, \, a \in \mathbb{R}, \qquad (15.10)$$

and write for $n$ large enough,

$$x_{n,a} = \cos(\theta_{n,a}) \text{ where } \theta_{n,a} \in (0, \pi). \qquad (15.11)$$

Here for $n \geq 1$, $|x| \leq 1 - n^{-2/3}$, Theorem 5.1 gives

$$\widetilde{K}_n(x,x) \sim n\sqrt{1 - x^2}.$$

Then

$$\sqrt{\frac{1 - x_{n,a}^2}{1 - x^2}} = 1 + O\left(\frac{|x - x_{n,a}|}{1 - x^2}\right)$$

$$= 1 + O\left(\frac{|a|}{n\left(1 - x^2\right)^{3/2}}\right)$$

$$= 1 + O\left(n^{-1 + \frac{3}{2}\tau_1}\right) = 1 + O\left(n^{-\tau_1}\right), \tag{15.12}$$

provided $|x| \le 1 - n^{-\tau_1}$ and $\tau_1$ is small enough. Also then

$$\theta_{n,a} - \theta_{n,0}$$

$$= \arccos\left(x_{n,a}\right) - \arccos\left(x_{n,0}\right)$$

$$= -\frac{1 + o\left(1\right)}{\sqrt{1 - x^2}}\left(x_{n,a} - x\right) = -\frac{\left(1 + o\left(1\right)\right)a}{\widetilde{K}_n\left(x, x\right)\sqrt{1 - x^2}}. \tag{15.13}$$

Recall from Theorem 13.4 that

$$\frac{\gamma_{n,n-1}}{\gamma_{n,n}} = \frac{1}{2} + O\left(n^{-C}\right). \tag{15.14}$$

Also let

$$\psi_{n,m}\left(x\right) = p_{n,m}\left(x\right)e^{-nQ_n(x)}\left(1 - x^2\right)^{1/4}. \tag{15.15}$$

Recall that if $g_n$ is defined by (13.9), and $x = \cos\theta$, we can recast (13.4) as

$$\psi_{n,m}\left(x\right) = \sqrt{\frac{2}{\pi}}\cos\left(\left(m - n\right)\theta + n\pi g_n\left(x\right) - \frac{\pi}{2}\right) + O\left(n^{-\tau_1}\right)$$

$$= \sqrt{\frac{2}{\pi}}\sin\left(\left(m - n\right)\theta + n\pi g_n\left(x\right)\right) + O\left(n^{-\tau_1}\right), \tag{15.16}$$

for $|x| \le 1 - n^{-\tau_1}$. From the Christoffel-Darboux formula, for $a \ne b$,

$$\Gamma_n := \frac{\widetilde{K}_n\left(x + \frac{a}{\widetilde{K}_n(x,x)}, x + \frac{b}{\widetilde{K}_n(x,x)}\right)}{\widetilde{K}_n\left(x, x\right)}\left(1 - x_{n,a}^2\right)^{1/4}\left(1 - x_{n,b}^2\right)^{1/4}$$

$$= \frac{\gamma_{n,n-1}}{\gamma_{n,n}}\frac{\psi_{n,n}\left(x_{n,a}\right)\psi_{n,n-1}\left(x_{n,b}\right) - \psi_{n,n}\left(x_{n,b}\right)\psi_{n,n-1}\left(x_{n,a}\right)}{a - b}. \tag{15.17}$$

Then using (15.14), (15.16),

$$
(\pi (a - b)) \, \Gamma_n
$$

$$
= \frac{\pi}{2} \left[ \psi_{n,n} (x_{n,a}) \, \psi_{n,n-1} (x_{n,b}) - \psi_{n,n} (x_{n,b}) \, \psi_{n,n-1} (x_{n,a}) \right] + O \left( n^{-C} \right)
$$

$$
= \sin (n\pi g_n (x_{n,a})) \sin (n\pi g_n (x_{n,b}) - \theta_{n,b})
$$

$$
\quad - \sin (n\pi g_n (x_{n,b})) \sin (n\pi g_n (x_{n,a}) - \theta_{n,a}) + O (n^{-\tau_1})
$$

$$
= \sin (n\pi g_n (x_{n,a})) \sin (n\pi g_n (x_{n,b})) \cos \theta_{n,b}
$$

$$
\quad - \sin (n\pi g_n (x_{n,a})) \cos (n\pi g_n (x_{n,b})) \sin \theta_{n,b}
$$

$$
\quad - \sin (n\pi g_n (x_{n,b})) \sin (n\pi g_n (x_{n,a})) \cos \theta_{n,a}
$$

$$
\quad + \sin (n\pi g_n (x_{n,b})) \cos (n\pi g_n (x_{n,a})) \sin \theta_{n,a} + O (n^{-\tau_1})
$$

$$
= \sin (n\pi g_n (x_{n,a})) \sin (n\pi g_n (x_{n,b})) (\cos \theta_{n,b} - \cos \theta_{n,a})
$$

$$
\quad - \sin (n\pi g_n (x_{n,a})) \cos (n\pi g_n (x_{n,b})) (\sin \theta_{n,b} - \sin \theta_{n,a})
$$

$$
\quad + (\sin \theta_{n,a}) (\sin (n\pi g_n (x_{n,b})) \cos (n\pi g_n (x_{n,a}))
$$

$$
\quad - \sin (n\pi g_n (x_{n,a})) \cos (n\pi g_n (x_{n,b}))) + O (n^{-\tau_1})
$$

$$
= (\sin \theta_{n,a}) \sin (n\pi [g_n (x_{n,b}) - g_n (x_{n,a})]) + O (|\theta_{n,a} - \theta_{n,b}|) + O (n^{-\tau_1})
$$

$$
= \sqrt{1 - x^2} \sin (n\pi [g_n (x_{n,b}) - g_n (x_{n,a})]) + O (n^{-\tau_1}), \tag{15.18}
$$

by (15.12), (15.13), provided $|x| \leq 1 - n^{-\tau_1}$ and $\tau_1$ is small enough. Next, from (13.9),

$$
n\pi [g_n (x_{n,b}) - g_n (x_{n,a})]
$$

$$
= n\pi \int_{x_{n,b}}^{x_{n,a}} \sigma_{Q_n} (t) \, dt + \frac{\theta_{n,b} - \theta_{n,a}}{2}
$$

$$
= n\pi \frac{a - b}{\widetilde{K}_n (x, x)} \left[ \sigma_{Q_n} (x) + O \left( |x_{n,a} - x_{n,b}|^\alpha \right) \right] + O (n^{-\tau_1})
$$

$$
= \pi (a - b) + O (n^{-\tau_1}),
$$

by (15.10), (15.13), the uniform Lipschitz condition (3.15) of $\{\sigma_{Q_n}\}$, and the asymptotic (13.7) for $\lambda_n = K_n^{-1}$. Combining this with (15.17), (15.18),

$$
\frac{\widetilde{K}_n \left( x + \frac{a}{K_n(x,x)}, x + \frac{b}{K_n(x,x)} \right)}{\widetilde{K}_n (x, x)} = \frac{\sin \pi (a - b)}{\pi (a - b)} + O \left( \frac{n^{-\tau_1}}{|a - b| \sqrt{1 - x^2}} \right).
$$

We then obtain (15.4) provided $a, b$ lie in a compact set and provided $|a - b| \geq n^{-\tau_1/2}$ and $|x| \leq 1 - n^{-\tau_1/2}$ for some small enough $\tau_1$. We turn to the case $|a - b| \leq n^{-\tau_1/2}$. From our bounds in Theorem 5.1 for $K_n$, and Cauchy-Schwarz, we have

$$\sup_{x,y \in I_n} |K_n(x,y)| e^{-nQ_n(x)-nQ_n(y)} \leq Cn.$$

Using Theorem 8.1, we have

$$\sup_{x \in [-1,1], y \in I_n} \left| \frac{\partial}{\partial x} \left[ K_n(x,y) e^{-nQ_n(x)} \right] e^{-nQ_n(y)} \right| \leq Cn^2.$$

Then for some $\xi$ between $x_{n,a}$ and $x_{n,b}$,

$$\left| \frac{\widetilde{K}_n(x_{n,a}, x_{n,b})}{\widetilde{K}_n(x,x)} - \frac{\widetilde{K}_n(x_{n,b}, x_{n,b})}{\widetilde{K}_n(x,x)} \right|$$

$$\leq \frac{|x_{n,a} - x_{n,b}|}{\widetilde{K}_n(x,x)} \left| \left[ \frac{\partial}{\partial x} K_n(x,y) e^{-nQ_n(x)} \right]_{|x=\xi, y=x_{n,b}} \right| e^{-nQ_n(x_{n,b})}$$

$$\leq C \frac{|a - b|}{\widetilde{K}_n(x,x)^2} n^2 \leq C \frac{n^{-\tau_1/2}}{1 - x^2} \leq n^{-\tau_1/4},$$

for $|x| \leq 1 - n^{-\tau_1/4}$, so using our asymptotic (13.7) for $K_n$,

$$\frac{\widetilde{K}_n(x_{n,a}, x_{n,b})}{\widetilde{K}_n(x,x)} = \frac{\sigma_{Q_n}(x_{n,b})}{\sigma_{Q_n}(x)} + O\left(n^{-\tau_1/4}\right),$$

and then again using our Lipschitz condition on $\sigma_{Q_n}$, we obtain

$$\frac{\widetilde{K}_n(x_{n,a}, x_{n,b})}{\widetilde{K}_n(x,x)} = 1 + O\left(n^{-\tau_1/4}\right)$$

$$= \frac{\sin \pi (b - a)}{\pi (b - a)} + O\left(n^{-\tau_1/4}\right).$$

So we have the general case of (15.4), with an appropriate choice of $\tau$ there. $\quad\square$

*Proof of Theorem 15.2.* We apply Theorem 2.3 in [30, p.11] with $\mu'_n = e^{-2nQ_n}$ on $I_n \subset \mathbb{R}$. The requirements there are the following four conditions. We simplify them to our situation, and note that we have proved above much more than what is needed to apply the results from [30]. The ordering of the zeros of $p_{n,n}$ is also different there.

**(I) Pointwise Asymptotics of Orthonormal Polynomials with a Rate**

Uniformly for $x = \cos\theta \in [-1+\varepsilon, 1-\varepsilon]$, and $m = n-1, n$,

$$p_{n,m}(x) \, e^{-nQ_n(x)} \left(1-x^2\right)^{1/4}$$

$$= \sqrt{\frac{2}{\pi}} \cos\left(\left(m-n+\frac{1}{2}\right)\theta + n\pi \int_x^1 \sigma_{Q_n} - \frac{\pi}{4}\right) + o\left((\log n)^{-1/2}\right),$$

where for some $C > 1, n \geq 1, x \in [-1+\varepsilon, 1-\varepsilon]$,

$$C^{-1} \leq \sigma_{Q_n}(x) \leq C.$$

We also need that the $\{\sigma_{Q_n}\}$ are equicontinuous in $[-1+\varepsilon, 1-\varepsilon]$, with

$$\omega(t) = \sup\left\{\left|\sigma_{Q_n}(x) - \sigma_{Q_n}(y)\right| : n \geq 1, x, y \in [-1+\varepsilon, 1-\varepsilon], |x-y| \leq t\right\}$$

$$= o\left(\left|\log t\right|^{-1/2}\right)$$

for $t \to 0+$.

**(II) Asymptotics of Leading Coefficients**

$$\frac{\gamma_{n,n-1}}{\gamma_{n,n}} = \frac{1}{2} + o(1), \, n \to \infty.$$

**(III) Asymptotic Spacing of Zeros**

Uniformly for $j, n$ with $x_{jn} \in [-1+\varepsilon, 1-\varepsilon]$,

$$\lim_{n\to\infty} n\sigma_{Q_n}\left(x_{j,n}\right)\left(x_{jn} - x_{j+1,n}\right) = 1.$$

**(IV) Asymptotics for Reproducing Kernels**

Uniformly for $x \in [-1+\varepsilon, 1-\varepsilon]$, we have

$$\lim_{n\to\infty} K_n(x,x) \, e^{-2nQ_n(x)} / \left(n\sigma_{Q_n}(x)\right) = 1.$$

Here (I) follows from Theorem 13.2 and (3.15). (II) follows from Theorem 13.4. (III) follows from Theorem 13.5. (IV) follows from Theorem 13.3.          □

For the proof of Theorem 15.3, we first establish some estimates for $\log|p_{n,n}|$:

**Lemma 15.4.** *(a) For $x \in [x_{nn}, x_{1n}]$,*

$$C_1 \log n + \log \min_{1 \leq j \leq n} \left|x - x_{jn}\right| \leq \log\left|p_{n,n}(x) \, e^{-nQ_n(x)}\right| \leq C_2 \log n. \qquad (15.19)$$

*(b) For* $x \in I_n \setminus [x_{nn}, x_{1n}]$,

$$C_3 n + \log \min_{1 \le j \le n} |x - x_{jn}| \le \log |p_{n,n}(x)| \le C_4 n \log (2 + |x|). \qquad (15.20)$$

*Proof.* (a) This follows easily from (14.4) and (14.8).
(b) Suppose that $x \in (x_{1n}, d_n)$. Since the zeros of $p'_{n,n}$ interlace those of $p_{n,n}$, both $p_{n,n}$ and $p'_{n,n}$ are increasing in $(x_{1n,\infty})$. Then for some $\xi$ between $x$ and $x_{1n}$,

$$\frac{p_{n,n}(x)}{x - x_{1n}} = p'_{n,n}(\xi) \ge p'_{n,n}(x_{1n})$$

so

$$\log |p_{n,n}(x)| \ge \log |x - x_{1n}| + \log \left| p'_{n,n}(x_{1n}) e^{-nQ_n(x_{1n})} \right| + nQ_n(x_{1,n})$$

$$\ge \log |x - x_{1n}| - C \log n + nQ_n(x_{1,n})$$

$$\ge \log |x - x_{1n}| + C_3 n,$$

by (14.5) and as $Q_{1,n}(x_{1n}) \sim 1$. Also as all zeros lie in $|x| \le 1 + \frac{c}{n}$ and by Theorem 13.1's asymptotics for $\gamma_{n,n}$,

$$\log |p_{n,n}(x)| \le \log \gamma_{n,n} + n \log (2 + |x|)$$

$$\le C_5 n + n \log (2 + |x|) \le C_4 n \log (2 + |x|).$$

$\square$

*Proof of Theorem 15.3.* Recall from (15.15) and (15.16) that

$$\psi_{n,n}(x) = p_{n,n}(x) e^{-nQ_n(x)} (1 - x^2)^{1/4} = \sqrt{\frac{2}{\pi}} \sin (n\pi g_n(x)) + O(n^{-\tau_1}),$$

where $g_n$ is given by (13.9). Recall too that we define $y_{jn}$ by $g_n(y_{jn}) = \frac{j}{n}, 1 \le j \le n$, and

$$g'_n(x) \sim -\sqrt{1 - x^2}, \quad |x| \le 1 - n^{-\tau_1}.$$

Let

$$I_{jn} = [y_{j+1,n}, y_{jn}]. \qquad (15.21)$$

By (13.12),

$$|I_{jn}| = y_{jn} - y_{j+1,n} \sim \frac{1}{n\sqrt{1 - y_{jn}^2}}. \tag{15.22}$$

Next, let

$$\mathcal{I}_n = \bigcup_{j=1}^{n} \left[ x_{jn} - n^{-100}, x_{jn} + n^{-100} \right].$$

Using Lemma 15.4(a), and our bounds in Theorem 7.1, we see that

$$\int_{\mathcal{I}_n} \left| p_{n,n} e^{-nQ_n} \right|^{\rho} \left| \log |p_{n,n}| \right|^{\tau} dx$$

$$\leq C \int_{\mathcal{I}_n} \left| 1 - x^2 \right|^{-\rho/4} \left( \left| \log \operatorname{dist} \left( x, \{x_{jn}\}_{j=1}^n \right) \right| + \log n + |nQ_n(x)| \right)^{\tau} dx$$

$$= o(n^{\tau}). \tag{15.23}$$

Next, consider an $I_{jn}$ with $|y_{jn}| \leq 1 - n^{-\tau_1}$. For $x \in I_{jn} \backslash \mathcal{I}_n$, we have

$$\left| \log |p_{n,n}(x)| \right| = \left| \log |\psi_{n,n}(x)| + nQ_n(x) - \frac{1}{4} \log (1 - x^2) \right|$$

$$= \left| nQ_n(y_{jn}) + O(\log n) + O(n^{1-\tau_1}) \right|,$$

using our uniform Lipschitz condition on $Q_n$. Using our spacing (15.22), we see that

$$\int_{I_{jn}\backslash\mathcal{I}_n} \left| p_{n,n} e^{-nQ_n} \right|^{\rho} \left| \log |p_{n,n}| \right|^{\tau} dx$$

$$= \int_{I_{jn}\backslash\mathcal{I}_n} |\psi_{n,n}(x)|^{\rho} \left| nQ_n(y_{jn}) + O(n^{1-\tau_1}) \right|^{\tau} (1 - x^2)^{-\rho/4} dx$$

$$= \left| nQ_n(y_{jn}) + O(n^{1-\tau_1}) \right|^{\tau} \left( \frac{2}{\pi} \right)^{\rho/2}$$

$$\times \int_{I_{jn}\backslash\mathcal{I}_n} |\sin(n\pi g_n(x)) + O(n^{-\tau_1})|^{\rho} (1 - x^2)^{-\rho/4} dx$$

$$= \left( (1 - y_{jn}^2)^{-\rho/4} + O(n^{-\tau_1}) \right) \left| nQ_n(y_{jn}) + O(n^{1-\tau_1}) \right|^{\tau} \left( \frac{2}{\pi} \right)^{\rho/2}$$

$$\times \left[ \int_{I_{jn}} |\sin(n\pi g_n(x)) + O(n^{-\tau_1})|^{\rho} dx + O(n^{-\tau_1}) \right]. \tag{15.24}$$

Here if $g_n^{[-1]}$ is the inverse function of $g_n$,

$$\int_{I_{jn}} |\sin(n\pi g_n(x))|^\rho \, dx = \int_{j/n}^{(j+1)/n} |\sin n\pi t|^\rho \frac{dt}{\left|g_n'\left(g_n^{[-1]}(t)\right)\right|}.$$

Using (13.10) and our Lipschitz condition on $\sigma_{Q_n}$, for $t \in \left[\frac{j}{n}, \frac{j+1}{n}\right]$,

$$g_n'\left(g_n^{[-1]}(t)\right) = -\sigma_{Q_n}(y_{jn}) + O(n^{-\tau_1}),$$

so using a standard integral [16, p. 369, (3.621.1)] and

$$\int_{I_{jn}} |\sin(n\pi g_n(x))|^\rho \, dx = \frac{1+o(1)}{\sigma_{Q_n}(y_{jn})} \frac{1}{n\pi} \int_0^\pi |\sin t|^\rho \, dt$$

$$= \frac{1+o(1)}{\sigma_{Q_n}(y_{jn})} \frac{2^\rho}{n\pi} \frac{\Gamma\left(\frac{\rho+1}{2}\right)^2}{\Gamma(\rho+1)} = \frac{2^\rho}{\pi} \frac{\Gamma\left(\frac{\rho+1}{2}\right)^2}{\Gamma(\rho+1)} (y_{jn} - y_{j+1,n})(1+o(1)),$$

recall (13.48). Substituting in (15.24) gives

$$\frac{1}{n^\tau} \int_{I_{jn} \setminus \mathcal{I}_n} \left|P_{n,n} e^{-nQ_n}\right|^\rho \left|\log|P_{n,n}|\right|^\tau dx$$

$$= A(1 - y_{jn}^2)^{-\rho/4} \left|Q_n(y_{jn}) + O(n^{-\tau_1})\right|^\tau (y_{jn} - y_{j+1,n})(1+o(1))$$

$$= A \int_{I_{jn}} \frac{|Q_n(y)|^\tau}{(1-y^2)^{\rho/4}} \, dy \, (1+o(1)),$$

where

$$A = \left(\frac{2}{\pi}\right)^{\rho/2} \frac{2^\rho}{\pi} \frac{\Gamma\left(\frac{\rho+1}{2}\right)^2}{\Gamma(\rho+1)}.$$

Adding over $j$, and using (15.23), gives

$$\frac{1}{n^\tau} \int_{|t| \leq 1 - n^{-\tau_1}} \left|P_{n,n} e^{-nQ_n}\right|^\rho \left|\log|P_{n,n}|\right|^\tau dt$$

$$= A \int_{|t| \leq 1 - n^{-\tau_1}} \frac{|Q_n(y)|^\tau}{(1-y^2)^{\rho/4}} \, dy \, (1+o(1)).$$

Next, our bounds on $p_{n,n}$ and $\log|p_{n,n}|$ give, much as above,

$$\frac{1}{n^\tau}\int_{1-n^{-\tau_1}\leq|t|\leq 1+n^{-\tau_1}}\left|p_{n,n}e^{-nQ_n}\right|^\rho\left|\log|p_{n,n}|\right|^\tau dt$$

$$\leq C\int_{1-n^{-\tau_1}\leq|t|\leq 1+n^{-\tau_1}}\frac{1}{(1-t^2)^{\rho/4}}(\log n)^\tau dt = o(1).$$

Finally, we can use restricted range inequalities and (15.20) to bound the rest of the integral.                                                                    □

# References

1. G. Akemann, J. Baik, P. Di Francesco (editors), *The Oxford Handbook of Random Matrix Theory*, Oxford University Press, Oxford, 2011.
2. G. Anderson, A. Guionnet, O. Zeitouni, *An Introduction to Random Matrices*, Cambridge Studies in Advanced Mathematics, 118. Cambridge University Press, Cambridge, 2010.
3. A. I. Aptekarev, R. Khabibullin, *Asymptotic Expansions for Polynomials Orthogonal with respect to a Complex Non-Constant Weight Function*, Trans. Moscow Math. Soc., 68(2007), 1–37.
4. A. I. Aptekarev, J. S. Dehesa, A. Martinez-Finkelshtein, *Asymptotics of Orthogonal Polynomial's Entropy*, J. Comp. Appl. Math., 233(2010), 1355–1365.
5. J. Baik, L. Li, T. Kriecherbauer, K. McLaughlin, C. Tomei, *Proceedings of the Conference on Integrable Systems, Random Matrices and Applications*, Contemporary Mathematics, Vol. 458, Amer. Math. Soc., Providence, 2008.
6. B. Beckermann, A. Martinez-Finkelshtein, E. A. Rakhmanov, F. Wielonsky, *Asymptotic Upper Bounds for the Entropy of Orthogonal Polynomials in the Szegő Class*, J. Math. Phys., 45(2004), 4239–4254.
7. D. Dai, M. E. H. Ismail, X.-S. Wang, *Plancherel-Rotach Asymptotic Expansion for Some Polynomials from Indeterminate Moment Problems*, Constr Approx, 40(2014), 61–104.
8. J. S. Dehesa, V. S. Buyarov, A. Martinez-Finkelshtein, E. B. Saff, *Asymptotics of the Information Entropy for Jacobi and Laguerre Polynomials with Varying Weights*, J. Approx. Theory, 99(1999), 153–166.
9. P. Deift, *Orthogonal Polynomials and Random Matrices: A Riemann-Hilbert Approach*, Courant Institute Lecture Notes, Vol. 3, New York University Press, New York, 1999.
10. P. Deift, D. Gioev, *Random Matrix Theory: Invariant Ensembles and Universality*, Courant Institute Lecture Notes, Vol. 18, New York University Press, New York, 2009.
11. P. Deift, T. Kriecherbauer, K. T-R McLaughlin, S. Venakides, X. Zhou, *Strong asymptotics of orthogonal polynomials with respect to exponential weights*, Comm. Pure Appl. Math. 52 (1999), 1491–1552.
12. P. Deift, T. Kriecherbauer, K. T-R McLaughlin, S. Venakides, X. Zhou, *Uniform asymptotics for polynomials orthogonal with respect to varying exponential weights and applications to universality questions in random matrix theory*, Comm. Pure Appl. Math. 52 (1999), 1335–1425.
13. P. J. Forrester, *Log-gases and Random matrices*, Princeton University Press, Princeton, 2010.

© The Author(s) 2018
E. Levin, D.S. Lubinsky, *Bounds and Asymptotics for Orthogonal Polynomials for Varying Weights*, SpringerBriefs in Mathematics,
https://doi.org/10.1007/978-3-319-72947-3

14. J. S. Geronimo, D. Smith, and W. Van Assche, *Strong asymptotics for orthogonal polynomials with regularly and slowly varying recurrence coefficients*, J. Approx. Theory, 72(1993), 141–158.
15. J. S. Geronimo and W. Van Assche, *Relative asymptotics for orthogonal polynomials with unbounded recurrence coefficients*, J. Approx. Theory, 62, (1990), 47–69.
16. I. S. Gradshteyn and I. M. Ryzhik, *Tables of Integrals, Series and Products*, Academic Press, San Diego, 1979.
17. J. Gustavsson, *Gaussian Fluctuations of Eigenvalues in the GUE*, Annales Institut H. Poincare, 41(2005), 151–178.
18. A. V. Komlov, S. P. Suetin, *An Asymptotic Formula for Polynomials Orthonormal with respect to a Varying Weight*, Trans. Moscow Math. Soc., 73(2012), 139–159.
19. A. V. Komlov, S. P. Suetin, *An Asymptotic Formula for Polynomials Orthonormal with respect to a Varying Weight II*, Sbornik Mathematics, 205(2014), 1334–1356.
20. A. B. Kuijlaars, *Riemann-Hilbert Analysis for Orthogonal Polynomials*, (in) "Orthogonal Polynomials and Special Functions, Leuven 2002" (E. Koelink and W. Van Assche, eds.), Lect. Notes Math. 1817, Springer-Verlag, pp. 167–210.
21. A. B. Kuijlaars, K. T-R McLaughlin, *Riemann-Hilbert Analysis for Laguerre Polynomials with Large Negative Parameter*, Comput. Meth. Funct. Theory, 1(2001), 205–233.
22. A. Kuijlaars, K. T-R McLaughlin, W. Van Assche, and M. Vanlessen, *The Riemann-Hilbert approach to strong asymptotics for orthogonal polynomials on* [−1, 1], Advances in Mathematics, 188 (2004), 337–398.
23. A. B. Kuijlaars and M. Vanlessen, *Universality for Eigenvalue Correlations from the Modified Jacobi Unitary Ensemble*, International Mathematics Research Notices, 30(2002), 1575–1600.
24. Eli Levin and D. S. Lubinsky, *Christoffel Functions, Orthogonal Polynomials, and Nevai's Conjecture for Freud Weights*, Constr. Approx., 8(1992), 463–535.
25. Eli Levin and D. S. Lubinsky, *Orthogonal Polynomials for Exponential Weights*, Springer, New York, 2001.
26. Eli Levin and D. S. Lubinsky, *Asymptotics for Entropy Integrals Associated with Exponential Weights*, J. Comp. Appl. Math., 156(2003), 265–283.
27. Eli Levin and D. S. Lubinsky, *Universality Limits in the Bulk for Varying Measures*, Advances in Mathematics, 219(2008), 743–779.
28. D. S. Lubinsky, *A New Approach to Universality Limits involving Orthogonal Polynomials*, Annals of Mathematics, 170(2009), 915–939.
29. D. S. Lubinsky, *Bulk Universality Holds in Measure for Compactly Supported Measures*, J. d' Analyse de Mathematique, 116(2012), 219–253.
30. D. S. Lubinsky, *Gaussian Fluctuations of Eigenvalues of Random Hermitian Matrices Associated with Fixed and Varying Weights*, Random Matrices: Theory and Applications, 5(2016), No. 3, 1650009, 31 pages, 10.1142/S2010326316500009X .
31. D. S. Lubinsky, *An Update on Local Universality Limits for Correlation Functions generated by Unitary Ensembles*, SIGMA, 12(2016), 078, 36 pages.
32. D. S. Lubinsky, E. B. Saff, *Strong Asymptotics for Extremal Polynomials Associated with Weights on* $\mathbb{R}$, Springer Lecture Notes in Mathematics, Vol. 1305, Springer, Berlin, 1988.
33. G. Mastroianni, G. V. Milovanovic, *Interpolation Processes, Basic Theory and Applications*, Springer, New Berlin, 2008.
34. K. T.-R McLaughlin and P. D. Miller, *The $\bar{\partial}$ Steepest Descent Method and the Asymptotic Behavior of Polynomials Orthogonal on the Unit Circle with Fixed and Exponentially Varying Nonanalytic Weights*, Int. Math. Res. Notices, (2006), Article ID 48673, 78 pages.
35. K. T.-R McLaughlin and P. D. Miller, *The $\bar{\partial}$ Steepest Descent Method for Orthogonal Polynomials on the Real Line with Varying Weights*, Int. Math. Res. Notices, (2008), Article ID rnn075, 66 pages.
36. H. N. Mhaskar and E. B. Saff, *Extremal Problems for Polynomials with Exponential Weights*, Trans. Amer. Math. Soc., 285(1984), 203–234.

37. H. N. Mhaskar and E. B. Saff, *Where does the Sup Norm of a Weighted Polynomial Live?*, Constr. Approx., 1(1985), 71–91.
38. H. N. Mhaskar and E. B. Saff, *Where does the $L_p$ Norm of a Weighted Polynomial Live?*, Trans. Amer. Math. Soc., 303(1987), 109–124.
39. P. Nevai, *Orthogonal Polynomials*, Memoirs of the American Math. Soc., 18(1979), No. 213.
40. P. Nevai, *Geza Freud, Orthogonal Polynomials and Christoffel Functions: A Case Study*, J. Approx. Theory, 48(1986), 3–167.
41. M. Plancherel and W. Rotach, *Sur les valeurs asymptotiques des polynomes d'Hermite* $H_n(x) = (-1)^n e^{x^2/2} \frac{d^n}{dx^n} \left( e^{-x^2/2} \right)$, Commentarii Mathematici Helvetici, 1(1929), 227–254.
42. E. A. Rakhmanov, *On Asymptotic Properties of Polynomials Orthogonal on the Real Axis*, Math. USSR. Sbornik, 47(1984), 155–193.
43. E. A. Rakhmanov, *Strong Asymptotics for Orthogonal Polynomials Associated with Exponential Weights on* $\mathbb{R}$, (in) Methods of Approximation Theory in Complex Analysis and Mathematical Physics, (eds. A. A. Gonchar and E. B. Saff), Nauka, Moscow, 1992, pp. 71–97.
44. E. B. Saff and V. Totik, *Logarithmic Potentials with External Fields*, Springer, New York, 1997.
45. B. Simon, *Two Extensions of Lubinsky's Universality Theorem*, J. d'Analyse Mathematique, 105 (2008), 345–362.
46. B. Simon, *Szegő's Theorem and its descendants*, Princeton University Press, Princeton, 2011.
47. G. Szegő, *A Hankel-féle formákról*, Mathematikai és Természettudomanyi Értesito, 36(1918), 497–538.
48. G. Szegő, *Über Orthogonalsysteme von Polynomen*, Mathematische Zeitschrift, 4(1919), 139–151.
49. G. Szegő, *Orthogonal Polynomials*, American Mathematical Society, Providence, 1939.
50. E. C. Titchmarsh, *Fourier Integrals*, Chelsea Publishing, New York, 1962
51. V. Totik, *Weighted Approximation with Varying Weight*, Springer Lecture Notes in Mathematics, Vol. 1569, Springer, Berlin, 1994.
52. V. Totik, *Asymptotics for Christoffel Functions with Varying Weights*, Advances in Applied Mathematics, 25(2000), 322–351.
53. V. Totik, *Universality and fine zero spacing on general sets*, Arkiv för Matematik, 47(2009), 361–391.
54. V. Totik, *Universality Under Szegő's Condition*, Canadian Math. Bulletin, 59 (2016), 211–224.
55. W. Van Assche, *Asymptotics for Orthogonal Polynomials*, Lecture Notes in Mathematics, Vol. 1265, Springer-Verlag, Berlin, 1987.
56. M. Vanlessen, *Strong Asymptotics of Laguerre-type Orthogonal Polynomials and Applications in Random Matrix Theory*, Constr. Approx., 25 (2007), 125–175.
57. Z. Wang and R. Wong, *Asymptotic expansions for second-order linear difference equations with a turning point*, Numer. Math., 94, (2003), 147–194.
58. Deng Zhang, *Gaussian Fluctuations of Eigenvalues in Log-gas Ensemble: Bulk Case I*, Acta Math. Sinica, 31(2015), 1487–1500.
59. Deng Zhang, *Gaussian Fluctuations of Eigenvalues in the Unitary Ensemble*, manuscript.
60. A. Zygmund, *Trigonometric Series*, Vols. 1,2, Second Paperback edition, Cambridge University Press, Cambridge, 1988.

# Index

© The Author(s) 2018
E. Levin, D.S. Lubinsky, *Bounds and Asymptotics for Orthogonal Polynomials for Varying Weights*, SpringerBriefs in Mathematics,
https://doi.org/10.1007/978-3-319-72947-3

Printed in the United States
By Bookmasters